Innovative Processing of Films and Nanocrystalline Powders

editor

Kwang-Leong Choy
Imperial College, UK

Innovative Processing of Films and Nanocrystalline Powders

Imperial College Press

Published by

Imperial College Press
57 Shelton Street
Covent Garden
London WC2H 9HE

Distributed by

World Scientific Publishing Co. Pte. Ltd.
P O Box 128, Farrer Road, Singapore 912805
USA office: Suite 1B, 1060 Main Street, River Edge, NJ 07661
UK office: 57 Shelton Street, Covent Garden, London WC2H 9HE

British Library Cataloguing-in-Publication Data
A catalogue record for this book is available from the British Library.

INNOVATIVE PROCESSING OF FILMS AND NANOCRYSTALLINE POWDERS

ISBN 1-86094-316-0

PREFACE

The processing and coating industries manufacturing advanced engineering ceramic films and powders are under increasing pressure to achieve lower processing costs, improved efficiency in production, better performance, and less pollution and this has provided an incentive and driving force for scientists and engineers all over the world to improve the present processes and develop novel methods. Their common goals are to lower processing costs and to manufacture new materials in a way which cannot be achieved by conventional methods. Information on newly developed processing methods is limited and difficult to obtain because it is scattered over the various research disciplines involved and so it is time to introduce a book to draw the information together and to review the latest research into these methods.

This book is based on a two-day international conference on "Innovative/Cost-Effective Materials Processing Methods" which was held at Imperial College on 6-7 July, 2000. The aim of the conference was to bring together international engineers and materials scientists to give technical presentations on the emerging innovative and cost-effective processing methods which are either in production or at the research and development stage. These methods are concerned with the fabrication of films and powders for structural and functional applications. The contributing authors have been chosen for their pioneering research on the innovative processing of films and powders and the methods described here have the potential to revolutionise materials fabrication by providing alternative cost-effective processing methods to increase both production efficiency and product performance.

The process principles and equipment used for each new method will be presented and in some cases examples of process modelling and advanced on-line *in-situ* process monitoring techniques will be included to help in understanding the process described. The issues and strategies associated with the scaling-up of the process will be addressed and various structural characterisations will be presented, together with an evaluation of the properties of a range of the materials concerned. Finally the correlation of processing, microstructure and properties, and the potential applications of each new method will be illustrated with specific case studies.

The innovative and cost-effective processing of films and nanocrystalline powders provides a useful source of reference for both academic and industrial researchers on the new processing methods which are emerging for the fabrication of these materials. In this book, as stated above, it is hoped that the benefits offered by these techniques will be clearly demonstrated, including their cost-effectiveness, efficiency of production, and the development of materials with unique microstructures for improved performance in engineering components for structural and functional applications. It will thus stimulate scientific and technological interest and lead to the eventual exploitation and implementation of these techniques in materials processing. Finally, it is envisaged that these materials processing methods have the potential to produce new advanced materials with novel microstructures and functions to meet the demand for the industrial and social requirements of the 21st century.

Kwang-Leong CHOY

LIST OF CONTRIBUTORS

Numbers in parenthesis indicate the pages on which the author's contribution begins.

KWANG-LEONG CHOY (1,15)
Department of Materials, Imperial College, London, United Kingdom.

JEAN-PIERRE SÉNATEUR *et al* (71)
LMGP, ENS de Physique de Grenoble, INPG, UMR CNRS 5628, BP 46, 38402 Saint Martin d'Hères Cedex, France.

STANISLAW MITURA (107)
Thin Films Division, Institute of Materials Science, Technical University of Lodz 90-924 Lodz, ul. Stefanowskiego 1/15, Poland.

ANDREW T. HUNT (147)
MicroCoating Technologies, Inc. 3901 Green Industrial Way, Chamblee, GA 30341, USA.

STEVEN L. GIRSHICK *et al* (165)
Department of Mechanical Engineering, University of Minnesota, Minneapolis, Minnesota, USA.

SOTIRIS E. PRATSINIS *et al* (193)
Institute of Process Engineering, ETH Zürich, CH-8092 Zürich, Switzerland.

DIETER VOLLATH *et al* (219)
Forschungszentrum Karlsruhe, Institut für Materialforschung III, P.O.Box 3640, D-76021 Karlsruhe, Germany.

JAMES F. GROVES *et al* (253)
University of Virginia, Thornton Hall A127, Materials Science and Engineering Department, Charlottesville, Virginia 22904-4240, USA.

CONTENTS

CHAPTER ONE

REVIEW OF ADVANCES IN PROCESSING METHODS: FILMS AND NANOCRYSTALLINE POWDERS

K. L. Choy

Department of Materials, Imperial College of Science, Technology and Medicine, Prince Consort Road, London SW7 2BP, UK

There is a rapid growth in the use of advanced engineering ceramic films and powders for structural and functional applications. The processing and coating industries manufacturing these engineering materials are under immense pressure to fulfil the targets of reducing processing costs and increasing production efficiency while achieving better product performance with lower associated pollution. Such demand has been brought about by the increased competitiveness of the processing industries in the global market which has prompted extensive research and development programmes around the world to develop innovative processing methods, which will be more cost-effective than conventional techniques.

The book highlights some of the innovative and cost-effective materials processing techniques relevant to the surface engineering and powder processing communities in particular. Their potential impact on cost-effectiveness, production efficiency and the generation of novel microstructures is also presented. This book is based on a two-day international conference on "Innovative/Cost-Effective Materials Processing Methods" which was held at Imperial College on 6-7 July 2000. The aim of the conference was to bring together international engineers and materials scientists to give reviews on the emerging innovative and cost-effective processing methods which are either already in production or at the research and development stage.

In order to appreciate the urgent need for new and econonmic materials processing methods, let us take a step back to look at the various conventional methods of producing coatings and powders, together with a summary of their advantages, limitations and their technological niche markets. These methods include solid state, wet chemical, liquid phase/molten state and vapour processing routes. There are microstructural differences between the materials manufactured by the various techniques and each method is particularly suited for particular materials system with specific shapes and volume, and for specific applications.

Solid-state processing routes such as the mechanical milling method involve mixing, grinding, calcination and sintering. These techniques are relatively simple but they tend to involve prolonged milling times and multiple cycles of processing which are also prone to contamination from the milling media and the powders can also suffer from chemical and phase inhomogeneities. Extensive milling is expensive, tedious and time consuming. Such a processing route is limited to the processing of powders and films, multilayer and functionally graded coatings cannot be produced in this way.

Wet chemical routes such as sol-gel, hydrothermal, sonochemical and electrodeposition methods, involve a large number of processing steps including mixing, chemical reactions, filtration, purification, drying, and calcination during the fabrication of powder and are therefore both tedious and liable to contamination. They also have poor chemical utilisation and there are environmental concerns over the waste treatment recycling of the chemical precursors and the disposal of by-products, especially during large-scale production. Moreover, the electrodeposition is limited to only conductive substrates. Sol-gel coated films tend to crack and there is a thickness limitation for each layer (about 1 μm) requiring that the process be repeated to obtain the necessary thickness. Hydrothermal methods are limited to powder processing and cannot be used to produce uniform films or coatings.

Other wet chemical routes such as slip casting, tape casting, calendering and screen printing may be appropriate for the fabrication of thick coatings on small areas of application but the considerable shrinkage associated with the removal of polymeric binders and plasticisers in subsequent sintering steps reduces the quality of the film in larger areas of fabrication.

Molten and semi-molten processing routes include methods such as gas atomisation, melt spinning and thermal spraying. Melt spinning tends to involve continuous ribbon for R&D rather than an industrially viable processing method. Thermal spraying and plasma spraying tend to produce a porous and 'splat-like' type of coating microstructure which limits the scope of application.

The vapour processing route seems to be the only route for producing very pure coatings with a good structural control at the atomic level or nanometer level, and also at low processing temperatures which is important, especially in the microelectronic and opto-electronic industries. This processing route is also suitable for the fabrication of ultrafine powders. There are two well established vapour processing routes, namely Chemical Vapour Deposition (CVD) and Physical Vapour Deposition (PVD) [e.g. evaporation, sputtering and ion plating].

PVD techniques have been widely used by physicists as they do not involve the complex chemistry and chemical reactions of the CVD-based techniques. The general characteristics of PVD are outlined in Table 1. There are hybrid processes such as reactive sputtering and reactive ion plating which introduce O_2, nitrogen or hydrocarbon for the deposition of oxides, nitrides and carbides (or DLC), but they still tend to have the limitations mentioned above.

Although CVD does not show similar limitations to PVD in Table 1, CVD-based methods generally involve complex chemistry of the precursors and the associated chemical reactions in the gas phase and heterogeneous reactions. These chemical issues are complex even for chemists and issues such as fluid flow, heat transfer and mass transport also need to be considered in the CVD process.

In view of the limitations of CVD and PVD techniques shown above, there is a need to develop novel or improved materials processing methods to offer simpler and/or cost-effective alternatives. This book is timely in addressing this issue and contains review articles on the emerging innovative and cost-effective vapour processing methods for thin films, thick coatings and nanocrystalline powders. These methods are either in production or at the research and development stage and have the potential to revolutionise existing materials fabrication methods by providing alternative methods to increase production efficiency and product performance.

Chapter 1 serves as an introductory chapter and provides the rationale for the book. It covers the background of various processing techniques, with emphasis on the vapour deposition techniques, their strengths and weaknesses and the need for improved or new processing methods. A brief description of various novel processing methods is highlighted and a comparison of these methods with their conventional counterparts is presented. Chapters 2 to 7 present the approach based on chemical vapour deposition methods. Chapter 2 describes the novel and cost-effective Electrostatic Spray Assisted Vapour Deposition (ESAVD) method for the fabrication of films and nanocrystalline powders. Chapter 3 shows the use of a new liquid precursor delivery and vapour generation method based on pulsed injection for the Metal Organic Chemical Vapour Deposition (MOCVD) of oxide thin films. The use of improved Radio Frequency Plasma Chemical Vapour Deposition (RFPCVD) for the synthesis of diamond particles and nanocrystalline diamond coatings is described in Chapter 4. Chapter 5 shows the use of a specially designed atomiser and flame synthesis in the Combustion Chemical Vapour Deposition

Table 1. A general comparison between CVD and PVD coating techniques, and their strengths and limitations

CVD	PVD
Sophisticated reactor and/or vacuum system *Simpler deposition rigs* with *no vacuum* system has been adopted in variants of CVD such as pyrosol, ESAVD, FAVD and CCVD.	*Sophisticated reactor and vacuum system.* Vacuum system is definitely required to facilitate the generation of vapour species and increase the mean free path of the vapour species to reach the substrate.
Expensive techniques for LPCVD, plasma assisted CVD, photo assisted CVD, MOCVD, ALEP, EVD. *Relatively low cost techniques* for AACVD and FACVD.	*Expensive techniques* for all types of PVD which include evaporation, sputtering and ion-plating
Non-line-of-sight process. It can • coat a complex shaped component • deposit coating with good conformal coverage	*Line-of-sight process.* There is • difficulty in coating a complex shaped component • conformal coverage is a problem
Tendency towards *volatile and/or toxic* chemical precursors. *Less volatile/more environmentally friendly* precursors have been adopted in variants of CVD such as ESAVD and CCVD.	Tendency for using *expensive sintered solid targets/sources* with difficulties for • large area deposition • varying the composition or stoichiometry of the deposits.
Multi-source precursors tend to produce non-stoichiometric film *Single source precursors* can overcome this problem.	*With multi targets or a single target* have difficulty in controlling the stoichiometry of the films because different elements will evaporate or sputter at different rates, except with the laser ablation method but this method is limited to R&D and not used for large scale/area deposition.
High deposition temperatures in conventional CVD. *Low to medium deposition temperatures* can be achieved using variants of CVD such as plasma assisted CVD, photo assisted CVD, MOCVD, ESAVD.	*Low to medium deposition temperatures*

(CCVD) method to deposit thin films. A new method of depositing nanocrystalline coatings using a hypersonic plasma particle deposition (HPPD) which involves the injection of gaseous reactants into a thermal plasma source is presented in Chapter 6. The synthesis of nanosized particles with well controlled characteristics using an aerosol-based flame microreactor is described in Chapter 7, and using the microwave plasma process in Chapter 8. The use of a new directed vapour deposition (DVD) to deposit coatings under a low vacuum gas flow environment compared with the conventional EBPVD, is described in Chapter 9. The process principles and characteristics of each method and the scope of its applications are briefly introduced and described as follows.

Chapter 2: ESAVD of films and nanocrystalline powders

Electrostatic Spray Assisted Vapour Deposition (ESAVD) is a simple and economically effective deposition method developed by Choy et al. This method involves spraying atomised precursor droplets across an electric field where the droplets undergo chemical reaction in the vapour phase near the vicinity of the heated substrate to produce a stable, solid film. Nanocrystalline powders can be produced using the ESAVD process by tailoring the homogeneous gas phase reaction. The ESAVD process can be performed in an open atmosphere without using any sophisticated reactor or vacuum system and is a rapid, simple and flexible process compared with the conventional vapour deposition processes such as CVD and PVD, and is also suitable for coating components of complex shapes. This method has a low consumption of chemical precursors and high deposition efficiency >(90%). The structure, stoichiometry, crystallinity and texture of films can be controlled by optimising the process parameters. It is a versatile technique, which can produce thin or thick films in the form of single layer, multilayer and compositionally gradient coating, and composite coating. ESAVD has been used successfully to deposit adherent thick films of Y_2O_3-ZrO_2 thermal barrier coatings (e.g. >250 μm) with the desirable columnar structure, combined with low thermal conductivity and thermal shock resistance, which are superior to those produced using the established commercial plasma spraying and electron beam PVD methods. Nanostructured thin films such as transparent and dense nanocrystalline anatase TiO_2 (crystalline size ~10nm) films have been deposited at 450°C, exhibiting a good optical transmission over a 370 nm wavelength. ESAVD is also suitable for depositing highly textured thin films on to amorphous substrates such as glass. The flexibility of ESAVD also allows the synthesis of simple oxides, multicomponent oxides, doped oxides, sulphides and selenides, and metallic [e.g. Pt, Pd, Ni] as well as polymeric materials [e.g. polyvinylidene fluoride] for both structural and functional applications which include catalytic coatings, solid oxide fuel cell components, ceramic membranes for selective gas separation, thermal barrier coatings, bioactive coatings, optical films and ferroelectric films for sensors and memory devices. Both conducting and non-conducting substrates can be used [e.g. metal, ceramic and plastics].

Chapter 3: Application of pulsed injection MOCVD to the deposition of oxide single layers and super lattices
At LMGP and Vilnius University, a new liquid precursor delivery and vapour generation method have been developed, based on pulsed injection for the Metal Organic Chemical Vapour Deposition (MOCVD) of oxide thin films. This method is based on fuel injection principles in thermal motors and involves a sequential injection of microamounts (e.g. few mg) of organometallic precursor solution into an evaporator, where flash volatilisation occurs. The thickness of the layer, coating stoichiometry and growth rate can be controlled precisely (digital growth) using pulsed injection MOCVD. This can be achieved by determining the number of drops injected (digital growth) for the fixed injection parameters (i.e. frequency, size of each drop), unlike the conventional MOCVD. Such unique capability is particularly important for the synthesis of superlattice and multilayers with complex stacking at nanometer scale level, or for the study of the effect of variation in stoichiometry on the properties of materials. For example: a $SrTiO_3$/ $YBa_2Cu_3O_7$ double layer, a crystallised Ta_2O_5/amorphous SiO_2 multilayer, and $YBa_2Cu_3O_7$/$PrBa_2Cu_3O_7$ and (La,Sr)MnO_3/$SrTiO_3$ superlattices with various modulation lengths, have each been deposited on single crystal substrates.

Chapter 4: Synthesis of nanocrystalline diamond coatings using improved Radio Frequency Plasma Chemical Vapour Deposition (RFPCVD)
Prof. S. Mitura of Lodz University, Poland reported the use of Radio Frequency Plasma Chemical Vapour Deposition (RFPCVD) for the synthesis of diamond particles and nanocrystalline diamond coatings. The effects of process parameters were investigated such as negative self-biased potential and hydrocarbon pressure on the substrate temperature and total energy transferred from plasma to the layer during deposition, and thus on the properties of the layers. (e.g. energy gap, optical absorption and refractive index and hardness). The use of RFPCVD of nanocrystalline diamond coatings for biomedical and cutting tool applications is presented here.

Chapter 5: Low cost Combustion Chemical Vapour Deposition of thin films in open atmosphere.

Dr. Hunt of Micro Coating Technologies Inc, U.S.A explained the state-of-the-art of a low cost thin film preparation method called Combustion Chemical Vapour Deposition (CCVD). This method combines the formation of submicron droplets using a specially designed atomiser (Nanomiser^{TM}), and flame synthesis to deposit thin films. During the CCVD process the chemical precursors are dissolved in an organic combustible solvent and the resulting solution is atomised into submicron droplets using Nanomiser^{TM}; the droplets are subsequently carried using a carrier gas into a flame where they undergo combustion and pyrolysis to deposit a thin film on to a substrate. The process can be conducted in open atmosphere and does not require expensive equipment. The CCVD process has some similar features to the flame synthesis of diamond but the CCVD process has been patented for applications other than diamond coatings and has been developed specifically for thin film deposition rather than thick film. A range of materials such as metal, ceramics and certain plastics (e.g. Teflon) can be used as substrates. CCVD has been used mostly for the deposition of oxide thin films (e.g. SiO_2, $Ba(Sr)TiO_3$, YSZ, LSC, PLZT etc), and for certain metallic films (e.g. Pt) which are less sensitive to oxygen for applications. The potential applications of CCVD for the fabrication of corrosion resistant, electronic and optical coatings are described.

Chapter 6: Hypersonic plasma particle deposition of nanocrystalline coatings

Prof. S. Girshick of the University of Minnesota, USA has described a new method of depositing nanocrystalline coatings using hypersonic plasma particle deposition (HPPD). The innovative feature of this process includes the use of gaseous reactants, which are injected into a thermal plasma source. The plasma is subsequently expanded through a nozzle, with the pressure dropping from slightly sub-atmospheric to about 2.7 mbars. The rapid cooling in the nozzle drives the nucleation of nanoparticles which are subsequently accelerated by the highly supersonic flow and deposited by high velocity inert impact to form a continuous film on a substrate placed perpendicular to the flow, slightly down-stream of the nozzle exit. HPPD can be used to deposit nanocrystalline films at rates comparable to conventional thermal spraying, and a few orders of magnitude higher than with CVD. The potential of this technique to produce wear- and friction-resistant coatings [e.g. SiC, TiC and layered Ti/TiC composites] is highlighted and the effect of processing conditions on the structure and mechanical properties of the nanostructured films is described. The typical grain sizes of the films are ~20nm, close to the particle sizes measured in the aerosol jet downstream of the nozzle. Such nanostructure produced by HPPD can increase the fracture toughness, decreasing modulus, friction coefficient and inherent flaw size, giving improved wear and friction properties in the ceramic protective coatings compared with coarse-grained coatings. The as-deposited SiC coatings (~20nm grain size) have a theoretical density of 95% and a hardness of 37 GPa (above the range of coarse-grained SiC by 26-36 GPa). Negligible grain growth was observed in these HPPD films on post-processing by vacuum sintering and hot isostatic pressing at temperatures <950°C. The deposition of layered metal-metal carbide composites [e.g. Ti/TiC] has also been demonstrated and could potentially increase the fracture toughness of the nanostructured films. The nanoparticles can be focused on a collimated beam by replacing the substrate with an aerodynamic lens. The potential of such modified HPPD for the fabrication of 3-D nanocrystalline microstructures such as micromechanical systems is described.

Chapter 7: Aerosol-based flame microreactor for the synthesis of nanoparticles
Wegner and Pratsinis from ETH, Zurich, reported the potential and flexibility of an aerosol-based flame microreactor for the inexpensive synthesis of nanoparticles with well controlled characteristics. A co-flow diffusion flame microreactor has been developed for the synthesis of nanoparticles and its performance has been tested for the synthesis of silica nanoparticles. Fumed silica nanoparticles were synthesised by oxidising the hexamethyldisiloxane (HMDSO) precursor in flame. The effect of the geometric dimension of a co-flow diffusion flame aerosol reactor on the properties of the powders was investigated and the influence of the oxidant flow rate and composition on the specific surface area of the powders was determined. The process of coagulation and sintering was investigated using thermophoretic sampling as a function of flame height. Nanoparticles with a specific surface area ranging from 30 to 240 m^2/gm, and of spherical to aggregated morphology, have been synthesised by controlling the temperature history of the process. The microreactor enhances reactant mixing in the flame and is more sensitive to changes in oxidant flow rates than a larger scale reactor operated with similar reactant flows. Such capabilities have broadened the scope of operations and given rise to greater flexibility in the quality of the product. Spherical particles can be synthesised in this way, as can high surface area, fractal-like particles, merely by changing the flow rate of the oxidant.

Chapter 8: Synthesis of nanopowders by the microwave plasma process - basic considerations and perspectives for the process on a larger scale.

Prof. D. Vollath of Karlsruhe University has developed a microwave plasma process capable of producing non-agglomerated nanosized powders in sufficient quantity for an industrial scale. Such powders can be fabricated by this process because the particles leave the plasma zone with electric charges of the same sign. This new process also allows the coating of nanosized particles in the second step of production by lowering the reaction zone temperature to as little as 150°C for ceramic, polymer or carbon coatings. Such coated particles can be used as starting materials for the fabrication of homogeneous composites to prevent interaction between particles. Both oxide and non- oxide (e.g. nitride, sulphide or selenide) nanoparticles (5-20nm) have been fabricated by the method and issues related to scaling up the process are addressed and a demonstration plant capable of producing batches of 100g described.

Chapter 9: Deposition of electron beam evaporant in a low vacuum gas flow environment

Dr. Groves of the University of Virginia, USA, describes the use of a new directed vapour deposition (DVD) to deposit coatings. In the DVD process, a carrier gas stream is passed over the surface of an electron beam-generated melt pool to modify the inherent vapour creation characteristics of the electron beam system, and influence the vapour atom diffusion on the deposition surface. The innovative features of DVD include low vacuum e-beam evaporation, a focused vapour stream and coupling plasma activation to enable rapid and efficient deposition of a coating, allowing variation in the coating microstructure and composition (e.g. multielement, multilayer and multistructured films) on substrates of many sizes and shapes. The DVD method exploits the best features of e-beam interaction with the material and adapts them to operation in a medium-to-low vacuum (1 microbar-10 millimibar) environment, whereas the conventional electron beam evaporation (EBPVD) requires a high vacuum (10^{-6}–10^{-4} mbar) environment. The avoidance of a high vacuum system in DVD can potentially reduce the capital equipment cost and so increase the cost-effectiveness of the coating process. The luminescence from the gas and vapour in the system allows the visual observation of atom transport and assists in precise manipulation of the process. The combination of visual observation and process modelling using Direct Simulation Monte Carlo would lead to the reproducible deposition of high performance coatings. The microstructural control of the DVD process could be enhanced by coupling plasma activation of the gas and vapour stream prior to deposition. Preliminary work showed that DVD might be suitable for the deposition of porous, columnar thermal barrier coatings for gas turbine applications, rather than the dense columnar structure at present produced by the EBPVD method.

The detailed description of each of these innovative methods and the associated deposition mechanism will be presented in subsequent chapters. In some cases, the process modelling and the use of advanced on-line and/or in-situ process monitoring techniques will be described. The issues and strategies associated with scaling up the processes will be addressed and the correlation of processing /microstructure/properties with the potential applications of each of the innovative methods will be illustrated with specific examples.

Table 2 outlines the special features of the various deposition methods and their distinct advantages over conventional techniques. The emerging low cost, non vacuum CVD-based techniques such as Electrostatic Spray Assisted Vapour Deposition (ESAVD) (Chapter 2) and Combustion Chemical Vapour Deposition (CCVD) (Chapter 5), offer the potential for scaling up for large area or mass production. These techniques open up new areas of application which might previously have been limited by the high cost and limitations of conventional CVD and PVD. ESAVD is a flexible method, which is capable of depositing thick films and nanocrystalline powders in addition to thin films, while Pulsed Injection MOCVD offers a unique and well controlled liquid evaporation method based on fuel injector principle for the precise growth of stoichiometric superlattices and multilayer thin films with complex stacking at nanometre scale, which is difficult to achieve using the conventional CVD and MOCVD methods.

Other recent innovative surface processing techniques such as Hypersonic Plasma Particle Deposition (HPPD) and Directed Vapour Deposition (DVD) offer precise and highly effective coating techniques to meet the increasing demand for better performance and higher quality in coated products. HPPD and DVD are not low-cost techniques as they involve high expenditure on capital equipment, however they do offer high deposition rates and improved coating quality and microstructure, which lowers the production time and the need for post-processing operations compared with thermal spraying and conventional EBPVD. They also have the potential to deposit novel microstructures with improved properties for niche applications where technological advantage can outweigh the processing cost, for example coatings produced by HPPD contain a nanocrystalline structure, where coatings produced by conventional thermal spraying methods contain a coarse-grained, 'splat-like' structure with micropores and microcracks. HPPD can produce coatings at a high deposition rate, a few orders of magnitude higher than in CVD, and with DVD a wider range of compositions and microstructures can be deposited such as multielement, multilayer and multistructured films. DVD operates at a low vacuum (1- 10 μbars) as opposed to EBPVD which requires high vacuum (10^{-6}–10^{-4} mbars). The use of a carrier gas to create and transport a focused vapour stream in the DVD process allows the deposition of coating substrates of different sizes and shapes which is more difficult to achieve using the current EBPVD method.

The Aerosol-based Flame Micro-reactor method for the precise synthesis of oxide nanoparticles with well controlled characteristics is also inexpensive. Although the Microwave Plasma Process is not a low cost method, it can produce high quality oxide, non-oxide nanoparticles, and coated nanoparticles without agglomeration at industrial scale. Such high quality nanoparticles will find applications in specialised niche markets which requires the materials in small quantities.

Summary and Outlook

The emerging innovative and cost-effective processing methods highlighted in this book are either in advanced production or at research and development stage. They have the potential to increase economic effectiveness and production efficiency, and to allow the fabrication of a broad range of materials to improve the performance of engineering components in both structural and functional applications. These processing methods also have the potential to produce new advanced materials with novel microstructures to meet the industrial and social requirements of the 21st century. It is hoped that this book will also provide a platform to air new ideas on materials processing. The synergies between the various processes presented here may lead on to the development of new and improved deposition techniques.

Table 2: Comparison of various emerging new deposition methods

a) **Method**	b) **Special features**	c) **Distinctive advantages**
Electrostatic Spray Assisted Vapour Deposition	• A variant of CVD method • Charged aerosol is generated using electrostatic spraying or electrostatic aerosol generator. • Open atmosphere processing method without using sophisticated reactor or vacuum system.	• Simple and low cost process • High deposition efficiency (>90%) • High deposition rate (1-10μm/min) • A shorter delivery time of the vapour precursor to the reaction zone, • Deposition of thin/thick films/ nanocrystalline powders.
Pulsed Injection MOCVD	• A variant of MOCVD method. • Uses single metalorganic liquid precursor source. • special fuel-like injector/evaporator for the sequential injection of microamounts (e.g. few mg) into an evaporator where flash volatilisation occurs.	• Fast evaporation of the precursor and a shorter delivery time of the vapour precursor to the reaction zone, and higher growth rate (μm/s); • Precise control of the coating thickness, stoichiometry and growth rate for the synthesis of superlattice and multilayer thin films with complex stacking at the nm scale level.
RadioFrequency Plasma CVD	• A variant of plasma CVD method • Deposited using dense plasma	• Deposition of thin films/nanocrystalline powders • This method is applied for a range of applications

Combustion Chemical Vapour Deposition	• A variant of flame synthesis method. • Submicron droplets are generated using a Nanomiser™. • Flame is used as heat source for combustion /pyrolysis reaction • Deposition process is performed in an open atmosphere	• Low cost method. • A thin film deposition process. • Fast evaporation of the precursor and a shorter delivery time of the vapour precursor to the reaction zone.
Hypersonic Plasma Particle Deposition	• A variant of thermal spraying and CVD. • Uses gaseous reactants. • A nozzle is used to expand the gaseous reactants in the plasma source and create the pressure /temperature drop for nucleation of nanoparticles. • Nanoparticles are accelerated by a highly supersonic flow. • Uses plasma to ionise and dissociate gases as well as to provide a substrate heating source. • Requires vacuum and more complex reactor for the generation of plasma.	• A method of depositing nanocrystalline coatings and 3-D nanocrystalline microstructures. • Lower deposition temperature and enhanced deposition rate.

Aerosol-based Flame Micro-reactor	• A variant of flame synthesis. • Uses gaseous precursors, and • A co-flow diffusion flame microreactor to enhance reactant mixing in the flame and it is more sensitive to changes in oxidant flow.	• An inexpensive method for the synthesis of oxide nanoparticles with well controlled characteristics. • Allows a wider processing window of operation. • A high flexibility in product quality. • Can produce oxide spherical particles & high surface area fractal-like particles by changing the oxidant flow rate.
Microwave Plasma Process	• A variant of plasma assisted CVD. • Capable of producing nanoparticles and coating the particles in the second stage of production to avoid agglomeration. • Deposition temperature can be lowered to 150°C for ceramic, polymer or carbon coating. • Requires vacuum and more complex reactor for the generation of plasma.	• Not a low cost method but it can produce high quality nanoparticles with very narrow particle size distribution and without agglomeration in sufficient quantity at industrial scale. • Both oxide and non oxide nanoparticles (5-20 nm).
Directed Vapour Deposition	• A variant of EBPVD • Uses a carrier gas to create and transport a focused vapour stream • Only low vacuum environment is required	• A rapid and efficient film deposition method. • Allows the deposition of coatings with a range of compositions and microstructures (e.g. multielement, multilayer and multistructured films). • Allows coating of substrates of different sizes and shapes.

CHAPTER TWO

PROCESS PRINCIPLES AND APPLICATIONS OF NOVEL AND COST-EFFECTIVE ESAVD BASED METHODS

K. L. Choy

Department of Materials, Imperial College of Science, Technology and Medicine, Prince Consort Road, London SW7 2BP, UK

1. Background

The use of advanced engineering ceramic films in structural and functional electronic applications is widespread and expanding rapidly. There is increasing demand for the surface coating and thin film providers to (1) reduce the production cost, (2) increase the deposition efficiency, and (3) improve the coating methods. Various techniques such as chemical vapour deposition (CVD), physical vapour deposition (PVD), plasma spraying and sol-gel have been investigated for the fabrication of thin films and coatings. The advantages and limitations of these conventional methods are discussed in Chapter 1.

Processing techniques based on the vapour phase deposition seem to be the only route capable of producing materials of high purity with a controlled microstructure and stoichiometry at the atomic level. CVD and PVD are the two best established commercial vapour deposition methods but it is generally agreed that CVD and PVD methods are too expensive for large units and mass production, and CVD often causes difficulties in controlling the stoichiometry of multicomponent films. These problems are associated with the differences in vapour pressure of the CVD reagents and the low deposition rate. Wheareas PVD is a line-of-sight process, it is difficult to coat substrates of complex shapes. PVD-based methods such as radio frequency sputtering give low deposition rates and low yield. CVD and PVD (apart from electron beam evaporation) of ceramic films are limited to thin films.

Although aerosol-assisted deposition methods have been used as low-cost fabrication techniques, they are used mainly for the fabrication of fine powders as has been noted in a review [1]. A few aerosol-assisted deposition methods have been used for the deposition of films with limited success, but they have not been applied to the fabrication of thick films. This is due to the lack of control of these methods, most of which are not true vapour deposition processes; also aerosol-based synthesis methods have relatively low deposition efficiency (less than 50% in most cases).

The lack of a suitably efficient and inexpensive vapour deposition method has prompted us to develop an alternative deposition method which will meet the demands of the surface coating industry.

2. Fundamental of ESAVD

A promising and emerging deposition technique called Electrostatic Spray Assisted Vapour Deposition (ESAVD) has been developed at Imperial College [2]. This method offers a significant reduction in production costs and is capable of both increasing the coating efficiency and providing a unique microstructure to enhance its performance.

2.1. Definition

Electrostatic Spray Assisted Vapour Deposition (ESAVD) is an aerosol-assisted chemical vapour deposition (AACVD) method (i.e. a variant of the CVD process). It involves the generation and charging of the aerosol which is then directed towards a heated environment where the charged droplets will undergo decomposition and chemical reaction in the vapour phase. The chemical reactions can be tailored to occur in different zones in order to fabricate dense films, porous coatings or nanocrystalline powders [3].

(a) Nanocrystalline powders can be produced by tailoring the chemical reaction to occur in the gas phase (e.g. homogeneous gas phase reaction, Zone C in Fig. 1). The powder can be collected either by the thermophoretic method or by using an electrostatic precipitator.

(b) A stable solid film with good coating adhesion can be deposited in a single production run by tailoring the chemical reaction to occur in or near the vicinity of the heated substrate (Zone D in Fig. 1).

(c) Porous films can be deposited by tailoring a combination of homogeneous and heterogeneous reactions.

The deposition occuring in Zones C and D corresponds to processes IV and III respectively in the AACVD [1].

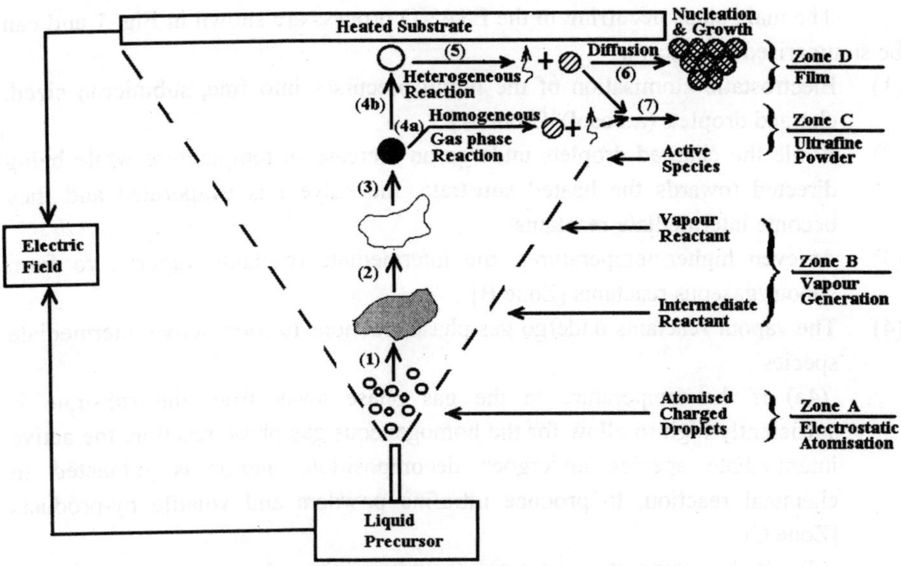

Fig.1. A schematic diagram of the ESAVD process.

A variant of ESAVD known as EAAJD (see section 3.4) has also been developed. The EAAJD uses an ultrasonic electrostatic generator for the aerosol which is subsequently passed through a charging unit. In the ESAVD process however, the aerosol is generated and charged simultaneously using electrostatic atomisation.

2.2. *Process principles*

The chemical precursor solution used in the ESAVD process can be prepared by dissolving solid or liquid starting chemicals in a solvent (e.g. an organic solvent with a high boiling point), or in a mixture of solvents to assist the vaporisation of the chemicals and to provide the additional thermal energy for the dissociation or decomposition of the chemicals. The generated aerosol from the electrostatic atomization is then delivered into a heated zone, where the solvent is rapidly evaporated and the intimately mixed chemical precursors undergo subsequent decomposition and/or chemical reaction in a heated environment, similar to the chemical vapour deposition reactions. The selection of appropriate chemical precursors for the decomposition and chemical reaction in the vapour phase is crucial to ensure that the process undergoes CVD reactions in order to provide the required precise structural control on a molecular scale.

The main steps occurring in the ESAVD process are shown in Fig. 1 and can be summarised as follows:

(1) Electrostatic atomisation of the liquid precursor into fine, submicron sized, charged droplets (aerosol) [Zone A].

(2) While the charged droplets undergo an increase in temperature while being directed towards the heated substrate, the solvent is evaporated and they become intermediate-reactants.

(3) At even higher temperatures, the intermediate reactants vaporise to form vapour/gaseous reactants [Zone B].

(4) The vapour reactants undergo gas phase reactions to form active intermediate species.

(4a) If the temperature in the gas phase away from the substrate is sufficiently high to allow for the homogeneous gas phase reaction, the active intermediate species undergoes decomposition, and/or is subjected to chemical reaction, to produce ultrafine powders and volatile by-products [Zone C].

(4b) If the temperature near the vicinity of the substrate is below that of decomposition of the active species, the active species is diffused across the boundary layer (i.e. a thin layer close to the substrate surface).

(5) The active species is subsequently absorbed on to the heated substrate surface, and a heterogeneous reaction occurs at the gas-solid (heated substrate) interface to produce the desired deposit and by-products [Zone D].

(6) The deposit is diffused along the heated substrate surface, forming the nuclei for subsequent growth of the film.

(7) The gaseous by-products are diffused away from the boundary layer of the substrate.

ESAVD is also a variant of the aerosol deposition process. However, unlike spray pyrolysis or other electrostatic deposition processes, in ESAVD-based processes it is essential that the chemical precursors undergo decomposition and/or chemical reactions in the vapour phase. This, like the true chemical vapour deposition (CVD) process, ensures the deposition of uniform films with good adhesion and structural control at the molecular level. In short, the ESAVD based methods are performed through processes C and D (as shown in Fig. 1) which are equivalent to Process IV and III in the aerosol deposition (see Fig. 2), and not Processes I and II which are typical processes for spray pyrolysis or other electrostatic deposition as described in reference [1].

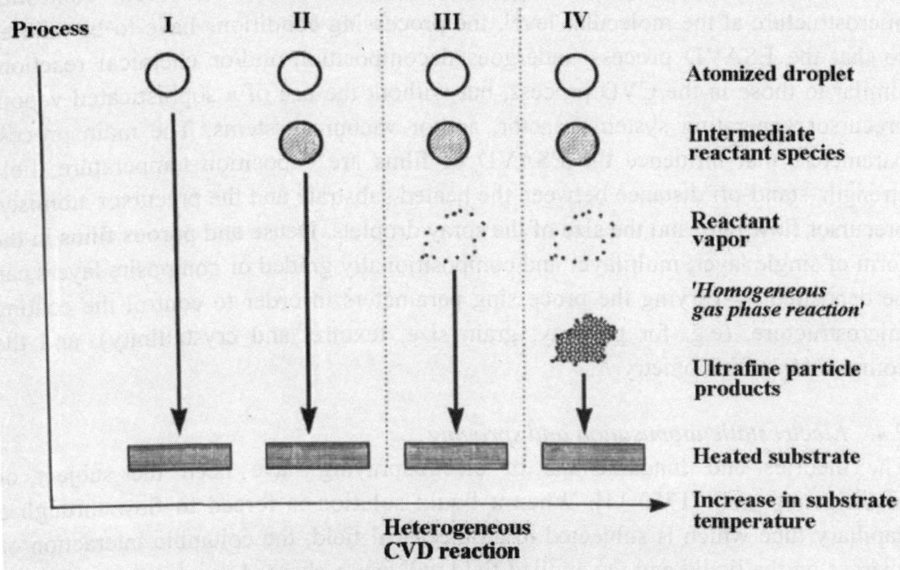

Fig. 2. Various aerosol deposition processes as a function of substrate temperature.

Processes I and II can be summarised briefly here. In Process I, the aerosol precursor droplets are sprayed directly on to a heated substrate, the solvent is subsequently removed through evaporation and decomposition of the precursor to produce the finished product. The deposited film may occasionally be subjected to a further sintering step to achieve a dense crystalline film. This process can produce dense thin films(<1 μm) with ultrafine crystalline structure, but the process has to be repeated several times in order to obtain the required thickness for thick film, making it a time-consuming process for the deposition of thick film. Also, as the film thickness increases, the deposited film tends to be porous, and cracking or spalling of the film can occur because of repeated drying, decomposition, and/or sintering procedures, whereas in Process II, the solvent is evaporated prior to its arrival at the substrate surface, and the precursor precipitate is subsequently deposited on to the heated substrate and decomposed and/or undergoes chemical reactions to produce the desired materials. This process has problems where uniform and dense films are required to be deposited.

2.3. Process parameters

In order to obtain high quality and adherent films with well controlled microstructure at the molecular level, the processing conditions have to be tailored so that the ESAVD process undergoes decomposition and/or chemical reactions similar to those in the CVD process, but without the use of a sophisticated vapour precursor generating system, reactor, and/or vacuum systems. The main process parameters that influence the ESAVD of films are deposition temperature, field strength, stand-off distance between the heated substrate and the precursor atomiser, precursor flow rate, and the size of the spray droplets. Dense and porous films in the form of single layer, multilayer and compositionally graded or composite layers can be deposited by varying the processing parameters in order to control the coating microstructure, (e.g. for porosity, grain size, texture, and crystallinity), and the composition/stoichiometry.

2.4. Electrostatic atomisation and spraying

The theories and fundamentals of electrospraying have been the subject of investigation since 1750 [4]. When a liquid solution is forced to flow through a capillary tube which is subjected to an electrical field, the columbic interaction of charges on the liquid and the applied field will cause charged droplets to be emitted from the capillary tube in different spraying modes as a result of different hydrodynamic mechanisms [5, 6]. The eletrospraying process has been applied commercially for paint spraying in the automobile industry and for insecticide spraying in the agricultural industry. The electrospray, especially in the cone jet mode has been adapted here for deposition using the ESAVD process and briefly reviewed in reference [3].

The electrostatic atomisation in the ESAVD process involves applying an electrical potential to a cylindrical spray nozzle which causes the formation of a stable cone with an attached jet. The cone is formed when the surface tension of the liquid precursor placed under an electric field balances with the electrical force and the hydrodynamic force [7]. Fernández de la Mora and Loscertales [8] have proposed the equations below (1) to estimate the droplet diameter and the current emitted from the liquid cone.

$$d_d = b_1(\varepsilon_r)\left(\frac{Q\varepsilon_r\varepsilon_0}{K}\right)^{1/3}, \qquad I = b_2(\varepsilon_r)\left(\frac{\gamma Q K}{\varepsilon_r}\right)^{1/2}, \tag{1}$$

where, d_d is droplet diameter (m), Q is liquid flow rate ($m^3 s^{-1}$), K is conductivity ($S \cdot m^{-1}$), γ is surface tension ($N \cdot m^{-1}$), ε_r is the relative permittivity of the liquid, ε_0 is permittivity of a vacuum ($C \cdot V^{-1} m^{-1}$), I is current (A), b_1 and b_2 are functions of the liquid permittivity. These relations are often called the scaling laws for electrohydrodynamic atomisation in the cone-jet mode and are only valid assuming that the liquid has a flat radial velocity profile in the jet.

The equation (1) are applicable only when the liquid has high conductivity and viscosity. If it does not, the equations (2) and (3) proposed by Gañán-Calvo et al [9] are more appropriate:

$$\text{if} \quad \sqrt[3]{\frac{\rho\varepsilon_0\gamma^2}{K\mu^3}} < 1, \quad \text{then} \quad d_j = \left(\frac{Q\varepsilon_0\sqrt{\beta - 1}}{K}\right)^{1/3} \tag{2}$$

$$\text{if} \quad \sqrt[3]{\frac{\rho\varepsilon_0\gamma^2}{K\mu^3}} > 1, \quad \text{then} \quad d_j = \left(\frac{\rho\varepsilon_0 Q^3}{\gamma K}\right)^{1/6} \tag{3}$$

where ρ is density, γ is surface tension ($N \cdot m^{-1}$), β is permitivity, K is conductivity ($S \cdot m^{-1}$), μ is viscosity, Q is liquid flow rate.

It is essential to establish the nature of the spraying mode produced during the electrostatic atomisation of the ESAVD process. A 'High Spec' imager and droplet size analyser (e.g. Malvern Mastersizer) have been used for the on-line monitoring of the spray pattern (i.e. modes of electrospraying and droplet size, respectively. In general a continuous flow of liquid through the meniscus generates spraying modes which includes pulsating (A), simple-jet (B), single cone-jet (C), multi-jet (D) and discharging mode (E) [10]. The spraying mode is strongly dependent on the applied electric field strength and is then transformed from A through B, C, D and E as the applied field strength increases. The multi-jet tends to occur at a higher field strength or when using a highly conductive precursor. This spray mode is unstable and produce a spray with a wide range of droplet sizes which will affect the nucleation and growth and lead to coatings with poor adhesion. The use of a very high electric field, is undesirable as it could cause the electrical discharge, and lead to irregular growth and result in a coating of powdery nature. The single cone-jet mode is the preferred spray mode for deposition as it offers a uniform, stable and continuous flow pattern with a narrow droplet size distribution. The preferred single cone-jet spray mode for the ESAVD process was obtained, for example at an electric field strength of between 5 - 10 x 10^6 V/m at flow rates of 10-25 ml/h. Although increase in the flow rate increases the area of the single cone jet mode [10], according to the scaling law [11] the liquid droplet size is proportional to the flow rate so that a flow rate below 25 ml/hr *(with a particular equipment configuration)* is preferred in order to obtain a spray with fine droplets. Large droplets may not decompose and/or react to form the finished product when they arrive at the heated substrate surface, and additional heat treatment is required to cause the reaction and transformation to the desired phase and densification of the deposit. In addition to field strength and precursor flow rate, the droplet size and modes of spraying can influence the properties of the fluid (e.g. its electrical conductivity, viscosity, dielectric constant, etc). The trajectories of the charged droplets under a different electrical field and substrate geometry are currently being investigated by modelling the electric field strength-trajectory of the spray droplets.

2.5. Electrical field

The electrical field is the driving force for the aerosol droplets moving up towards the substrate. The trajectories of the droplets are dependent on the electrical field, the temperature and velocity of the air, and the space between the atomiser and substrate. A 3-D model based on Gauss' law [see equation (4)] has been developed to simulate the electrical field in the ESAVD deposition rig,

$$-\nabla(\varepsilon_i\, \varepsilon_o \nabla V) = \rho \;, \qquad (4)$$

where V is the electrical potential, ε_i is the relative permittivity for air, liquid precursor and coating, ε_0 is the permittivity of vacuum and ρ is the space charge density. The effect of ρ on the electrical field is negligible in this case and can be assumed to be zero in the calculation. The FEMLAB software from COMSOLAB, Sweden, has been used to solve Equation (4) in the geometrical configuration of the planar deposition rig.

Figs. 3 and 4 show the electrical displacement vectors in an ESAVD rig when (a) there is no coating on the grounded metal substrate; and (b) the substrate is coated with a 30 μm thick dielectrical layer. The growth of a dielectric coating on the substrate surface changes the direction and concentration of electrical field. Fig. 3(b) and 4(b) show that the electrical field is now being more diverted away from the substrate coated with a dielectric layer, as compared with the case where there is no initial dielectric layer [see Fig. 3(a) and 4(a)]. Such additional diversion reduces the delivery of the aerosol to the substrate which will result in the decrease in the deposition rate. Further modelling work is in progress to investigate the effects of process parameters on the electrical field, droplets trajectories and hence the coating microstructure and deposition rate.

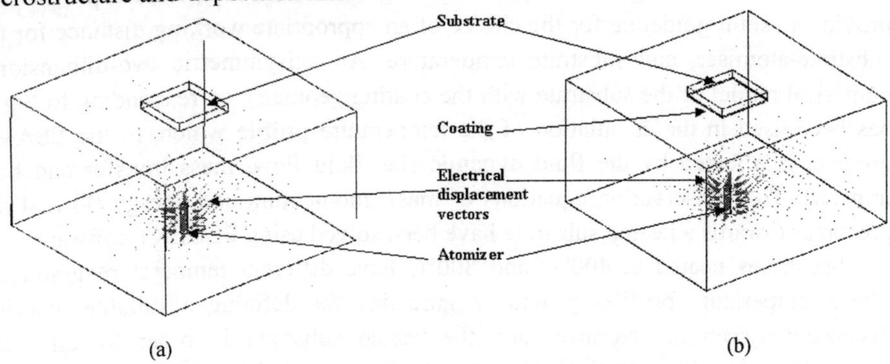

Fig. 3. Electrical field displacement vectors in a 3-D view for (a) an uncoated substrate and (b) a substrate coated with a dielectric layer.

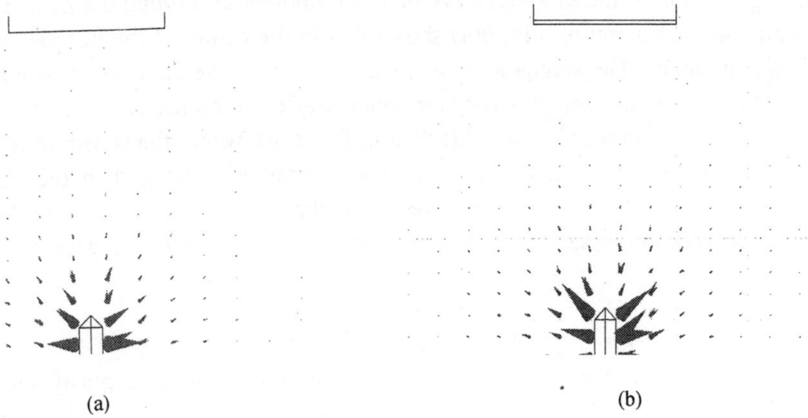

Fig. 4. A magnified Y-Z projection of electrical field displacement vectors in the space between the atomiser and the (a) uncoated, and (b) coated substrates.

2.6. *Thermal field*

Substrate temperature and temperature distribution are important aspects of the ESAVD process and both the decomposition of the precursor and the kinetics of the deposition are strongly dependant on these factors. The presence of a temperature gradient between the heated substrate and the atomiser causes precursor decomposition and nucleation/growth of the films could occur at various different locations (see Fig. 1). The substrate temperature has to be controlled in order to tailor the heterogeneous reactions to take place near the heated substrate surface. The theoretical investigation of the thermal gradient and temperature profile will provide a useful guideline for the choice of an appropriate working distance for the substrate-atomiser, and substrate temperature. An axisymmetric two-dimensional numerical model of the substrate with the coating geometry corresponding to Fig. 1 has been used in the calculation of the temperature profile which, in the ESAVD process, is affected by the fluid dynamic (i.e. fluid flow, mass transfer and heat transfer). The conservation equations of mass, momentum and energy flow of the precursor towards a heated substrate have been solved using FLUENT software.

Substrates heated at 400°C and 800°C have different temperature gradients. These temperature profiles provide a guideline for defining a suitable working distance between the atomiser and the heated substrate in order to tailor the appropriate chemical reactions to occur for the deposition of films or powders. Substrate heated to a higher temperature has a higher temperature gradient as shown in Fig. 5. This is due to a higher velocity of ambient air around the heated substrate at a higher temperature. Fig. 6(a) shows that in the centre of the hot plate there is a 'stagnant zone'. The stagnant zone is the zone where the air velocity is almost zero and the aerosol movement is not significantly affected by the convection flows. This zone is slightly bigger in Fig. 6(a) than in Fig. 6(b), where the substrate is heated to a higher temperature. However, there is no dramatic changed in the size of the stagnant zone. These results provide a guideline on identifying a suitable position for the aerosol atomiser in the stagnant zone during the deposition process.

2.7. Microstructure of ESAVD deposits

ESAVD-based processes can synthesise materials with high purity and precise structural control at the nanometer scale level, at the relatively low processing temperatures which are suitable for the fabrication of nanocrystalline materials. Fig. 7 shows that we can nano-engineer the ESAVD deposition to produce nanocrystalline powder (10-20nm), nanoporous material and nanostructured thin films (crystallites of less than 10 nm) of Eu-Y_2O_3, by varying the deposition conditions in order to tailor the chemical reaction so that it occurs in the appropriate deposition zone (as described in section 2.1).

15

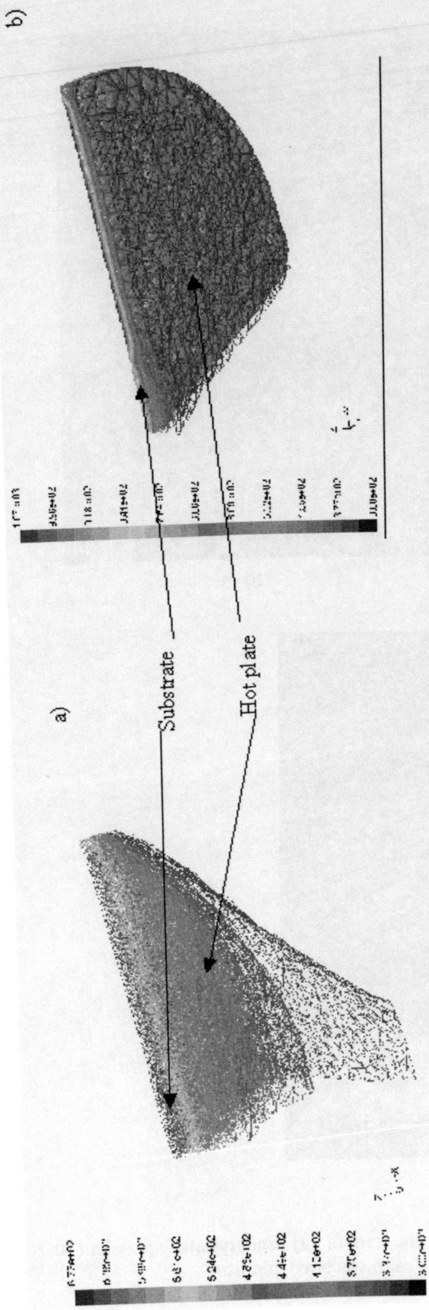

Fig.5. A diagonal view of a section of the temperature profile of a substrate heated to (a) 673K, and (b) 1073K

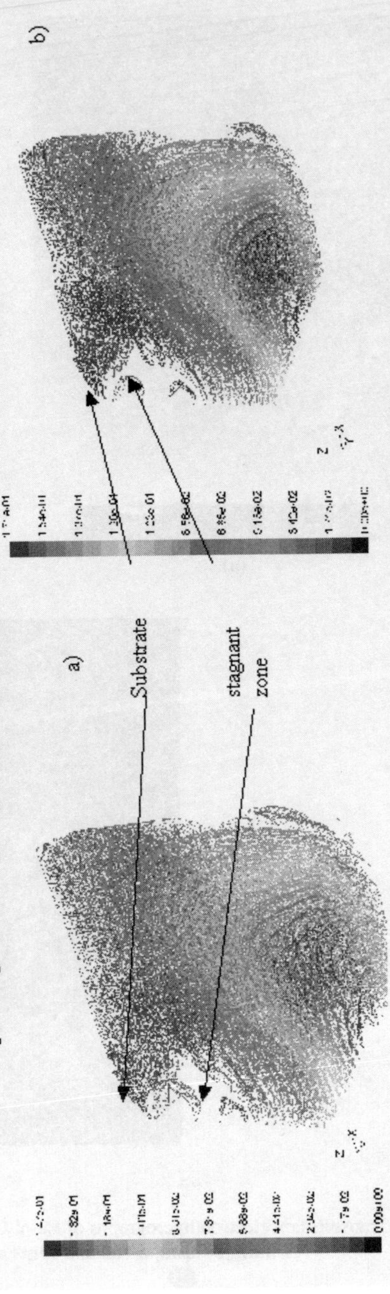

Fig 6. The velocity magnitude of ambient air when the substrate is heated to (a) 673 K and (b) 1073K.

(a) (b)

(c)

Fig. 7. Transmission electron micrographs of Eu-Y_2O_3 in the form of (a) nanocrystalline powder (10-20 nm), (b) nanoporous thin film and (c) nanostructured thin film.

2.7.1. Nanocrystalline powders

For the synthesis of nanocrystalline powders, the deposition temperature is tailored to be sufficient to initiate and maintain the homogeneous reaction in the gas phase (see Zone C in Fig. 1) which will lead to the nucleation of solid from the gas phase. Such homogeneous nucleation normally occurs at high deposition temperatures and highly concentrated reactive species. Apart from the synthesis of nanoparticles of Eu-Y_2O_3, Y_2O_3-ZrO_2 (YSZ) is further example of nanocrystalline powders which has been synthesised using the ESAVD process. The nanocrystalline YSZ powders (mean particle size 15 nm) synthesised by the ESAVD method at a temperature of 500°C using a mixture of zirconium and yttrium alkoxide precursor, have been shown to reduce the sintering temperature of YSZ to 950°C compared with the 1400°C required by the commercial YSZ powders (e.g. Tosoh, Japan) to achieve full density [13].

2.7.2. Film

The nucleation and growth of the deposits produce either thin or thick film with either dense or porous microstructure. The nucleation and growth kinetics are influenced by the deposition temperature and concentration of reactive species, which in turn influence the microstructure and thus the properties of the coatings.

(a) Porous films

Porous films have potential applications as catalysts, sensors, filters etc. These films can be deposited by tailoring a combination of homogeneous and heterogeneous gas phase reactions and the pore size, pore volume and surface area can be controlled by varying the processing conditions to reflect the degree of homogeneous and heterogenous reactions. The use of ESAVD to synthesise porous anode and cathode films for solid oxide fuel cell applications is described in section 3.1.3.

(b) Dense films

The deposition of dense films can be achieved by tailoring heterogeneous chemical reaction to occur in the vicinity of the heated substrate surface. This results in the adsorption of mobile atoms/monomers on the substrate surface, which diffuse to the preferred sites on the substrate surface to form stable nuclei. The growth occurs by the subsequent addition of adsorbed monomers to sites with the lowest free energy [e.g. kinks and ledges to form crystallites] and on coalescence a thin crystal is formed. The heterogeneous reaction can lead to various microstructures depending on the successive corresponding two-dimensional growth. For the ESAVD process, the types of growth generally observed are amorphous, polycrystalline [e.g. columnar, equaixed] and epitaxial. These growth features have been also observed in other vapour deposition processes such as CVD and PVD processes showing that there are similarities of the coating microstructures between CVD [3], PVD [14, 15] and ESAVD processes. The amorphous film is normally deposited using a low substrate temperature, which limits the mobility of the absorbed species and prevents the species from migrating to low free energy sites where nucleation could occur.

The nucleation of polycrystalline coatings normally occurs at different sites on the substrate surface, leading to the growth of islands which subsequently coalesce to form a polycrystalline layer. The nature of the surface and the absorbed species strongly influence the nucleation process. The control of polycrystalline growth, in particular the size and nature of the crystallites, is important to obtain the desired properties in the ESAVD films. This type of growth can lead to a columnar or equiaxed structure. Columnar structure consists of columnar grains of preferred orientation that grow from the first nucleated, equiaxed layer deposited on the substrate. Such microstructure is deposited at a low substrate temperature using a high concentration of reactive species, which results in limited diffusion. This type of microstructure is desirable for high temperature structural applications, e.g. thermal barrier coatings where the stress-strain columnar structure can accommodate the high temperature thermal cycling (see section 3.1.1). However an equiaxed structure consists of fine and randomly orientated grains which is desirable for applications requiring high mechanical strength and fracture toughness. Such microstructure can be deposited using a low substrate temperature but an even higher concentration of reactive species during the ESAVD process. Such process conditions allow a rapid supply of reactants to re-nucleate new grains, and the lower substrate temperature limits the surface diffusion and restrains the migration of monomers to preferred crystal lattice sites.

Epitaxial or highly textured growth occurs where the deposited crystals maintain the same crystallographic orientation as the substrate. Such films can be deposited using a high substrate temperature but low concentration of reactive species. This allows a rapid surface diffusion and leads to the nucleation and growth of solids in a particular preferred orientation, with lower interfacial free energy than other orientations. This type of growth is illustrated using the ESAVD of ZnS (see section 3.2.2).

In order to ensure the deposition of dense films, homogeneous nucleation must be prevented because any powder particles nucleated from the gas phase can be deposited onto the substrate surface causing interruption of the surface nucleation and growth which would lead to a powdery coating with irregular structure and poor coating adhesion. ESAVD allows control and variation of the microstructure of the coating by varying the process parameters. The structure, stoichiometry, crystallinity and texture of films can be controlled by optimising process parameters such as substrate temperature and precursor composition. The effect of these parameters on the microstructure of coatings will be illustrated using a case study on ESAVD of CdS films (see section 3.2.2). Examples of various films and coatings and their microstructures are illustrated in the following case studies.

2.8. *Advantages*

The main advantages of ESAVD-based processes over other deposition techniques are as follows.

(i) The ESAVD process can be performed in open atmosphere for the deposition of oxide and some less oxygen sensitive non-oxide materials without the need for any sophisticated reactor and/or expensive vacuum system, unlike CVD or PVD techniques so that the fabrication cost can be significantly reduced.

(ii) Under optimum deposition conditions, ESAVD is a true CVD process with the non-line-of-sight feature and it can therefore be used to coat complex shaped engineering components and to deposit coatings with good conformal coverage, unlike PVD which is a line-of-sight process which poses problems when coating components of complex shapes.

(iii) Materials of high purity can be produced at relatively low deposition temperatures (a fraction of their melting point), for example cubic YSZ can be deposited at 550°C.

(iv) It has a high deposition efficiency (>90%) because the precursor is directed to the substrate under the electric field which minimises loss of precursor to the surroundings, whereas deposition efficiency with CVD, PVD and thermal spraying tends to be less than 50%.

(v) It has a low consumption of chemical precursors and uses a relatively low concentration of precursor (e.g. 0.05M).

(vi) It tends to require a single-source precursor which provides good molecular mixing of chemical precursors and enables the synthesis of multicomponent materials with a well controlled stoichiometry. Such stoichiometric films are difficult to achieve by CVD and PVD (except laser ablation, for R&D only) techniques because CVD tends to use different reagents with different vaporising pressures, and PVD techniques such as sputtering tend to use a multitarget, giving different sputtering rates.

ESAVD simplifies the vapour precursor generation and delivery method compared with the conventional CVD method which uses a special bubbler/vaporiser method which lowers the cost of the deposition process.

The ESAVD-based method does not involve the use of the scrubber or effluent gas handling system to remove any unreacted precursors or toxic by-products as in necessary in CVD because the starting materials are relatively more environmentally friendly and only a small quantity (e.g. 0.05 M or less) is used. The process has a higher precursor conversion rate and deposition efficiency (above 90%).

It is a versatile technique which can produce thin or thick films in the form of a single layer, multilayer, compositionally gradient or composite coating, as well as nanosized powders.

It is a simple and flexible technique which allows the incorporation of the deposition technique into on-line processing, and the process can be scaled up for a large area and mass production.

ESAVD also has distinct advantages over other low-cost thin film or thick film processing techniques. For example, ESAVD is a one-step process unlike the sol-gel technique which has a thickness limitation of about 1 μm for each layer, and the process has to be repeated to obtain the required film thickness and also sol-gel fabricated films tend to crack during subsequent sintering. Other conventional ceramic thick film fabrication techniques such as tape casting, slip casting, screen printing and calendering may be appropriate for small area configurations, but the large shrinkage associated with the removal of polymeric binders and plasticisers in subsequent sintering steps reduces the film quality in large area fabrication.

2.9 Comparison of ESAVD with other electrospraying techniques

The comparison of the ESAVD-based method with other electrospraying techniques has been reviewed and discussed in [1, 3] and will be briefly summarised and highlighted here. The theoretical and experimental studies of the behaviour of the liquid jet in an electric field had been initiated since 19th century [4, 7, 16]. The industrial applications of electrostatic spraying were found in electropaint spraying for cars and insecticide spraying in the agricultural industry, however electrostatic spraying has only been adopted for materials fabrication since the 1980s. Different technical names [2, 17-24] such as electrostatic spray-assisted vapour deposition, electrostatic aerosol-assisted jet deposition, corona spray pyrolysis, electrostatic spray pyrolysis, electrostatic spray deposition, electrospray organometallic chemical vapour deposition and gas-aerosol reactive electrostatic deposition have been used by different research groups to illustrate electrospray-based processing, as reviewed by Choy [1].

Although electrospraying has been used in these deposition processes, the actual deposition process and mechanism in each case is rather different because of the nature of the chemical precursor used, the deposition conditions and the design of the processing equipment. These differences result in different deposition mechanisms (i.e. Process I-IV in Fig. 2) and hence products of different quality. Some of the more successful techniques are proprietary and subject to patents.

In general, the research activities on electrospray processing have focused on the ultrafine powders [e.g. corona spray pyrolysis [17], Electrostatic Spray Pyrolysis [18], and quantum dot composites [e.g. electrospray organometallic chemical vapour deposition [19], gas-aerosol reactive electrostatic deposition [21]. However, there are very limited studies on the use of electrospray-based processing methods for the deposition of uniform and adherent thin or thick films with a well controlled microstructure. Schoonman et al [22, 23] reported the use of Electrostatic Spray Deposition for the deposition of $LiCoO_2$ and $LiMnO_4$ with particulate-like microstructure. This indicates a particulate deposition rather than the atomistic deposition which normally occurs in the CVD and PVD processes. Such microstructure may be adequate for an application such as a cathode on rechargeable lithium, but atomistic deposition is essential for the deposition of nanophase materials or uniform, dense and reproducible thin or thick films. The chemistry of the precursors and the process conditions have therefore to be tailored carefully to facilitate deposition by the Process III as depicted in Fig. 2 (i.e. equivalent to heterogeneous CVD reaction in Zone D in Fig.1) for the production of films, or via Process IV (Fig.2) for the fabrication of nanoparticles.

There was some work initiated in the late 1980s [24] at Imperial College using corona spray pyrolysis for the deposition of films by electrospraying a metalorganic compound on to a heated tube to generate ionised vapour. It was then delivered into a CVD reactor to cause the decomposition and chemical reaction to deposit semiconducting materials for the manufacture of an integrated circuit. The vapour precursor delivery system, enclosed reactor and special effluent gas handling system are however still required in this process, as in conventional CVD.

Choy and her team [2, 25] have since initiated and developed a simpler and more cost-effective ESAVD-based method for the synthesis of thin/thick films and nanocrystalline powders with a well controlled structure at the molecular or nanometer level which is reviewed and described in this chapter. The ESAVD-based methods can be scaled up for large area planar deposition or complex 3-D deposition by using a multiple nozzle and controlled movement of the engineering components and/or the electrostatic spray nozzle. A detailed understanding of the fundamental science underlying the deposition processes was needed and a strong interdisplinary R&D team consisting of materials scientists, chemists, pyhysicists, mathematicians and engineers has played an important role in the progress of the ESAVD-based processes and has enabled them to be used commercially for the fabrication of a wide range of thin and thick films in both structural and functional applications. Further information on the commercial applications can be found on the website www.imptl.com.

2.10. *Applications*

A variety of thin and thick coatings, with dense or porous microstructures, in the form of single layer, multilayer or compositionally graded coatings have been deposited using the ESAVD methods for a wide range of applications as shown in Fig. 8.

Simple oxides (e.g. TiO_2, Al_2O_3, and SiO_2), multicomponent oxides [e.g $PbTiO_3$, $BaZrO_3$, $La(Sr)MnO_3$, $YaBa_2Cu_3O_{7-x}$, CaO-P_2O_5-SiO_2], doped oxides (e.g. Y_2O_3-ZrO_2, Eu:Y_2O_3], sulphides and selenides [e.g. CdS, ZnS, CdSe] have been used, and metallic [e.g. Pt, Pd, Ni] and polymeric films [e.g. polyvinylidene fluoride] have been deposited using the ESAVD-based processes for both structural and functional applications. These include thermal barrier coatings, catalytic coatings, bioactive coatings, optical films, ferroelectric films for sensors and memory devices, solid oxide fuel cell components, bioactive coatings, and ceramic membranes for selective gas separation. This technique allows coatings to be deposited on to either conducting or non-conducting substrates [e.g. metal, ceramic and plastics] and nanocrystalline powders [e.g. YSZ, CdS, Eu:Y_2O_3] with narrow size distribution have also been deposited using these processes.

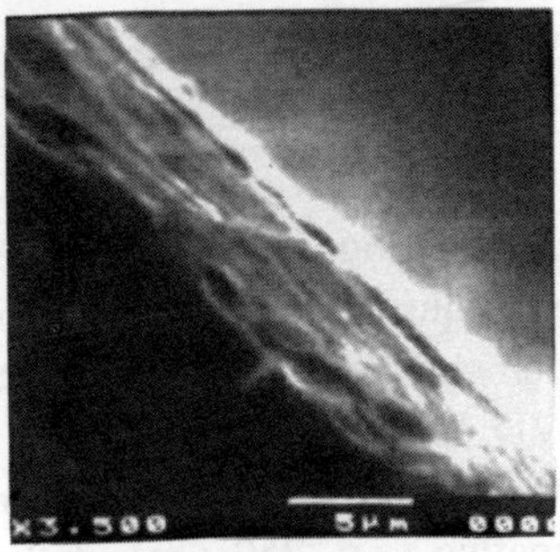

(a)

- Thermal barrier coating (YSZ)
- Tribology (YSZ/Al_2O_3)
- Industrial gas production (CGO, La-based oxides
- Piezo & pyroelectrics (PZT, PVDF)
- Solar cells (CdS, CdSe)
- Flat panel display (ITO)
- High Tc superconductor (YBCO)

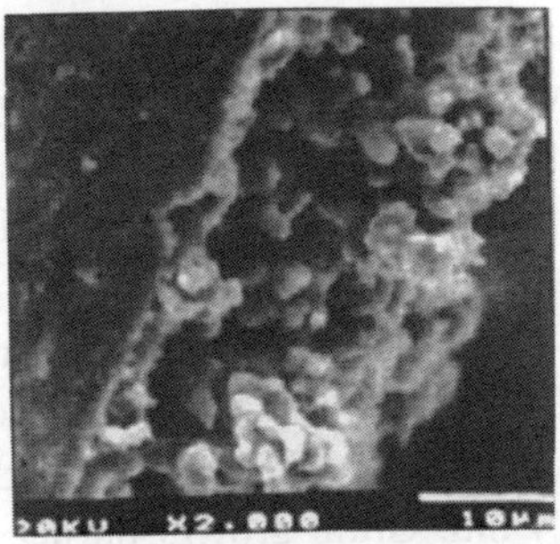

(b)

- Catalysts (Pd/Al_2O_3, Ni/Al_2O_3)
- Sensors (YSZ, SnO_2)
- Biomedical (Hydroxyapatite bioactive coating)

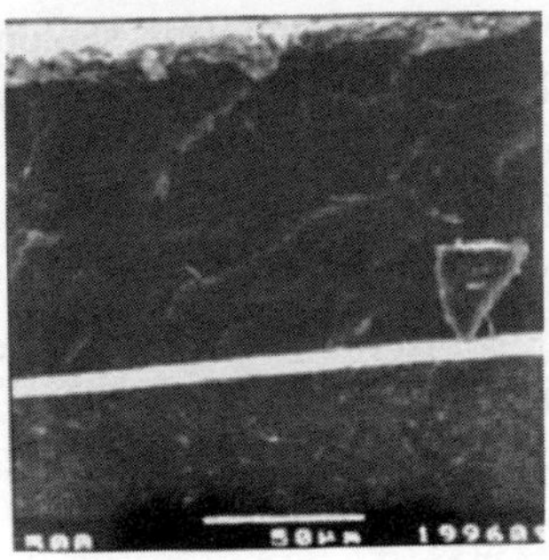

(c)

- Porous/dense/porous multilayers such as porous anode, dense electrolyte and porous anode for fuel cell application

Fig. 8. The versatility of the ESAVD process to deposit (a) dense, (b) porous and (c) multilayer coatings and their associated applications.

3. Case Studies

3.1. ESAVD of thick films

3.1.1. Thermal barrier coatings

ESAVD can also be used to synthesise thick films, and coatings with thicknesses of above 400□m have been deposited using this technique. One of the potential applications of thick coatings is the thermal barrier coatings (TBC) for turbine components. The non line-of-sight nature of ESAVD and its ability to tailor the microstructure of coatings make it appropriate for the coatings of 3-D engineering components. An example of ESAVD of a thick coating (e.g. 250 μm) of 8wt% Y_2O_3-ZrO_2 (YSZ) on a Ni-alloy substrate is illustrated in Fig. 9. The coating has been deposited using a mixture of alkoxide of zirconium and yttrium precursors in an alcohol solvent (e.g. butanol) at 550°C. The coating exhibits the desirable stress-strain columnar structure, and the tailored micropores and microcracks necessary to lower the thermal conductivity of the TBC. Such microstructure has been achieved by tailoring a combination of heterogeneous and homogeneous chemical reactions to occur in the vicinity of the heated substrate (see Fig. 1). The XRD (Fig. 10) shows the presence of a major tetragonal phase and minor cubic phase, however no monoclinic phase has been detected or confirmed by other Raman spectroscopy studies. The crystallite size of the as-deposited YSZ determined from the XRD data using the Scherrer formula was found to be about 4 nm. Grain growth during the heat treatment at 1000°C for 2 hours has increased the crystallite size tenfold.

The TBCs produced by ESAVD have been compared with those produced by other established commercial methods such as plasma spraying (for stationary turbine components) and the very expensive electron beam physical vapour deposition (for rotating turbine components such as turbine aerofoils). The comparisons showed that the thermal conductivity of the TBCs produced by ESAVD is lower than that of those produced by the EBPVD method, and is comparable to those produced by plasma spraying. The thermal shock resistance of the TBCs produced by ESAVD is as good as that of TBCs produced using the EBPVD method. ESAVD is therefore a potentially cost-effective method to produce TBCs for both stationary and rotating turbine components.

Fig. 9. Cross-section SEM of 8wt-% Y_2O_3-ZrO_2 coating on an Ni-based superalloy substrate.

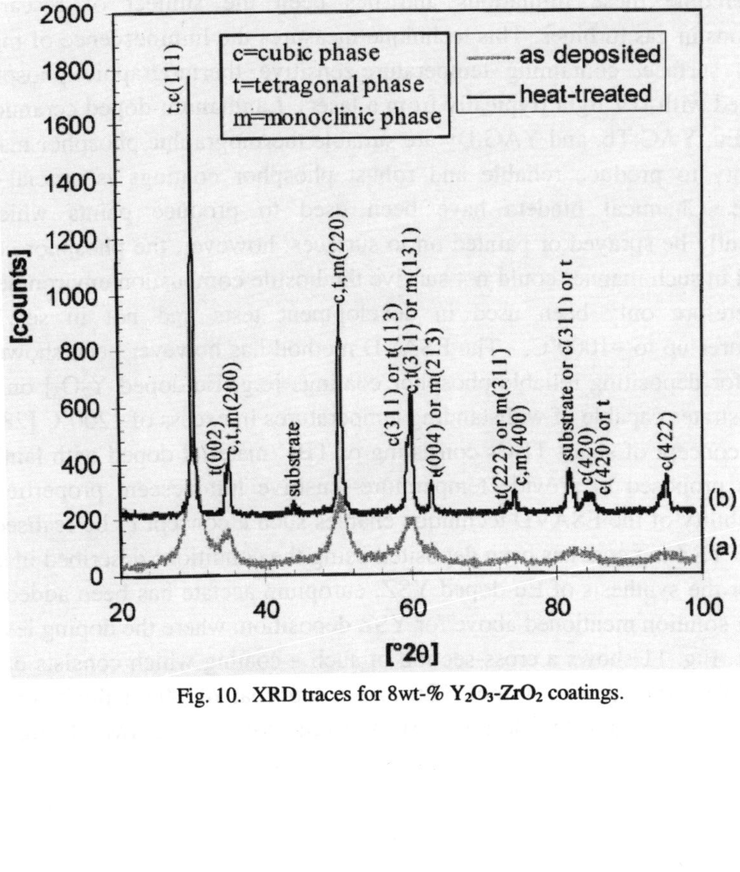

Fig. 10. XRD traces for 8wt-% Y_2O_3-ZrO_2 coatings.

3.1.2. Smart thermal barrier coatings

The flexibility and versatility of the ESAVD method can also facilitate materials development. For example, the development of smart thermal barrier coatings with both thermal insulation and instrument properties for temperature detection, and potentially also an erosion indicator etc. [27]. The design of the cooling system and the longevity of TBCs requires a good understanding of the surface temperature and heat flux. A reliable method of measuring temperatures and heat fluxes experienced by turbine components at high temperatures under realistic operating or service conditions would therefore be of great benefit in design and development. There are many problems and limitations with the current thermometry techniques in such high temperature environments. Thermocouples provide the measurement of localised temperature only, and are difficult to install on rotating components, and may not withstand such a severe environment. Infrared pyrometry raised problems associated with signal corruption from the flames, degradation of optical cleanliness and variations in surface emissions. The use of thermographic phosphor however may overcome these limitations and has been the subject of research for applications in gas turbines. This technique measures the luminescence of materials when its surface, containing temperature-sensitive thermographic phosphor, is illuminated with UV light (typically from a laser). Lanthanum-doped ceramics such as Y_2O_3:Eu, YAG:Tb, and YAG:Dy are suitable thermographic phosphor materials. The ability to produce reliable and robust phosphor coatings is crucial to the technique. Chemical binders have been used to produce paints which can subsequently be sprayed or painted on to surfaces, however, the phosphor coatings produced in such manner could not survive the hostile combustion environment and have therefore only been used in development tests and not in service, at temperatures up to ~1000°C. The ESAVD method has however been shown to be suitable for depositing reliable phosphor coatings [e.g. Eu-doped Y_2O_3] on to Ni-alloy substrates capable of withstanding temperatures in excess of 1200°C [28].

The concept of smart TBCs consisting of TBC material doped with lanthanide ions was proposed to provide temperature-sensitive luminescent properties [27]. The flexibility of the ESAVD technique enables such a concept to be realised [29]. An 8wt% YSZ top coat has been deposited using the conditions described in section 3.1.1. For the synthesis of Eu-doped YSZ, europium acetate has been added to the precursor solution mentioned above for YSZ deposition, where the doping level was 1 mol.-%. Fig. 11 shows a cross-section of such a coating which consists of 15μm thick of YSZ doped with 1 mol.%Eu and a top coat of 50μm thick YSZ. The microstructure exhibits the desirable stress-strain columnar growth features, and microporosity and microcracks tailored to low the thermal conductivity in TBCs. The XRD pattern was identified as a mixture of tetragonal and cubic phases similar to those 8 wt-%YSZ. Fig. 12 shows that both single (i.e. YSZ:Eu) and dual (i.e. YSZ:Eu/YSZ) coatings exhibit similar curves for temperature response of the time-decay constant. In each case the lifetime decay constant decreases slightly from room temperature to about 377°C and then decreases rapidly up to a temperature of 827°C. Such similarity indicates the repeatability and reliability of the results that may be obtained from a subsurface layer. Such a temperature-measuring technique is highly precise, the standard error on the individual data points of each set being estimated at less than 1%. The results have demonstrated the suitability of ESAVD to produce smart TBCs incorporating thermal insulation, and temperature-sensing properties. This offers a potential non-contact and *in-situ* temperature monitoring method for turbine components coated with suitable phosphor-based thermal barrier coatings. A small dopant level of 1 mol.-%Eu allows phosphorescence detectable from the surface or a subsurface layer which allows temperature measurements up to at least 827°C. Higher detection temperatures could be achieved using other doping materials.

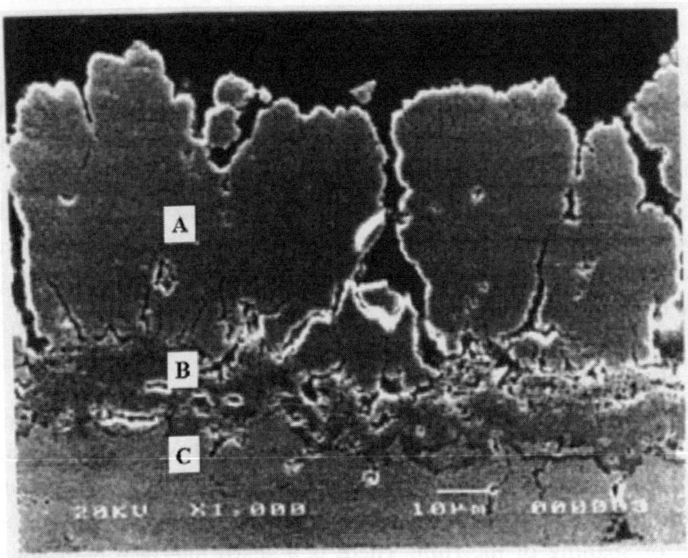

Fig. 11. Cross-section SEM of multilayer smart thermal barrier coatings deposited using ESAVD after annealing at 1050°C for 2h, where A: 8wt-% Y_2O_3-ZrO_2 coating; B: Y_2O_3-ZrO_2 doped with 1 mol.-%Eu coating; and C: Ni-based superalloy substrate.

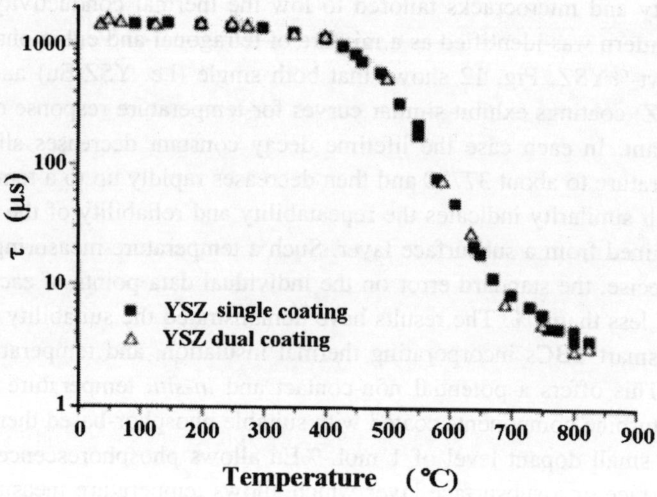

Fig. 12. The temperature response of the life time decay constant for single and dual coatings of YSZ:Eu.

3.1.3. Solid oxide fuel cells

ESAVD has been used successfully to deposit dense electrolytic and porous electrodes for solid oxide fuel cells (SOFCs) by varying the processing parameters in order to control microstructure, grain size, composition, surface area, pore size and particle size. The flexibility and versatility of the ESAVD process also makes possible the fabrication of SOFC components in the form of multilayer structures of positive electrode/electrolyte/negative electrode, compositionally graded electrodes, and also the development of improved cathode/electrolyte systems.

(a) Dense electrolytes

Fig. 13 [13] shows a dense and uniform 9mol% Y_2O_3-ZrO_2 (YSZ) electrolyte layer (65μm thick), deposited by ESAVD onto a porous La(Sr)MnO_3 (LSM) substrate at 650°C, with a deposition rate of 1 μm/min. The fracture surface of the coated substrate shows that the YSZ has good adhesion to the porous LSM substrate with no cracking or spalling of the film. There is a sharp interface between the dense YSZ and the porous LSM substrate and no undersirable interfacial reactions occur between the YSZ and LSM. The activation energy of the YSZ electrolyte layer determined from the Arrhenius plot of bulk conductivity versus temperature between 266°C and 600°C is 1.13 eV, which is in agreement with the bulk YSZ materials.

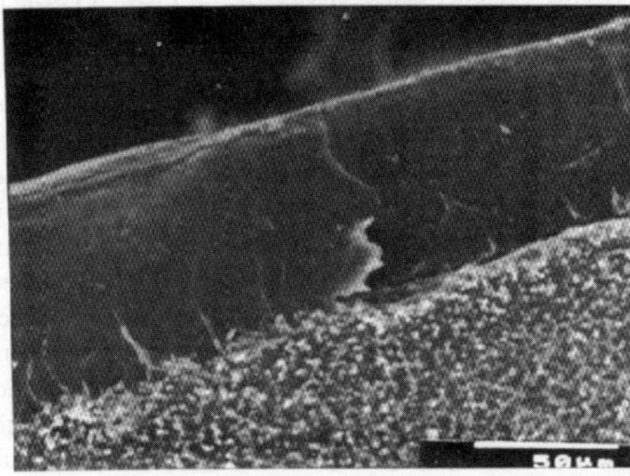

Fig. 13. Cross-section SEM of a dense 9 mol% Y_2O_3-ZrO_2 electrolyte layer deposited by ESAVD on to a porous La(Sr)MnO_3 substrate.

(b) Porous electrodes

A porous La(Sr)MnO_3 cathode layer (17 μm thick) with uniform pore size can also be deposited using ESAVD onto a YSZ substrate as shown in Fig. 8(b). The LSM and YSZ substrates used in Figs. 14 and 8(b) respectively, have been prepared using the conventional ceramic powder processing route. The degree of porosity in the LSM cathode films is strongly dependent on the substrate temperature and precursor flow rate. Porous LSM films are obtained when the precursor flow rate is greater than 0.1 ml/min, and substrate temperature is above 550°C which allows a combination of homogeneous and heterogeneous reactions to occur, whereas dense LSM films are obtained below such temperatures and precursor flow rates. The LSM films deposited at temperatures below 500°C have an amorphous or nanocrystalline structure. Highly crystalline LSM with the desired perovskite phase can be deposited at substrate temperatures above 650°C [30], which is still a lower deposition temperature than that of conventional ceramic sintering (1100-1200°C) of LSM. This high ceramic sintering temperature often leads to the formation of the undesirable La_2ZrO_3 arising from the interfacial reaction between the LSM layer and the YSZ electrolyte.

(c) Multilayer structures of positive electrode/electrolyte/negative electrode (PEN)
Using the ESAVD method, SOFCs with a multilayer structure of porous La(Sr)MnO_3 (LSM) cathode, dense electrolyte Zr(Y)O_{2-x} (YSZ), porous anode NiO-YSZ, have been fabricated on ferritic steel and porous ceramic substrates (5cm x 5cm) in a single production run by controlling deposition conditions and changing the precursor solutions. Fig. 8(c) shows an example of the multilayer LSM/YSZ/NiO-YSZ deposited onto a porous LSM-supporting substrate. It shows the uniform multilayer structure with a clear interface, no cracking or spalling is observed and the thickness of each layer can be varied. The feasibility of using ESAVD to produce SOFC components has the potential for manufacturing fuel cell components cost-effectively compared with the expensive electrochemical vapour deposition method (in excess of 10 million pounds), and the vacuum plasma spraying processing technology which have hindered somewhat the commercialisation of SOFC technology for efficient power generation. Although there are many groups trying to adapt low-cost conventional ceramic processing methods such as tape casting [31], screen printing [32], slurry coating [33] and colloidal method [34, 35] for the fabrication of PEN structures, while these conventional methods are appropriate for small area configurations, the large shrinkage associated with the removal of polymeric binders and plasticers in subsequent sintering steps reduces the quality of PEN assemblies which are larger in area. It can also be difficult to retain adequate porosity within the supporting electrode structures during the co-firing stage for densification of the thick film electrolyte.
(d) Compositionally graded electrodes
ESAVD has also been used to fabricate compositionally graded Ni-YSZ anode films. The development of this type of film has been prompted by the need to overcome or minimise the thermal expansion mismatch of Ni-YSZ cermet with the YSZ electrolyte, especially in the presence of high Ni content because the electrical conductivity of the Ni-YSZ anode cermet is dependent on this. The percolation threshold for the conductivity of the Ni-YSZ anode system is about 30 vol. % nickel. The coefficient of thermal expansion (CTE) of Ni-YSZ cermet increases linearly with Ni content so there are concerns about the thermal expansion mismatch between the anode and electrolyte, which can result in large stresses causing cracking or delamination of the anode films during fabrication and/or operation.

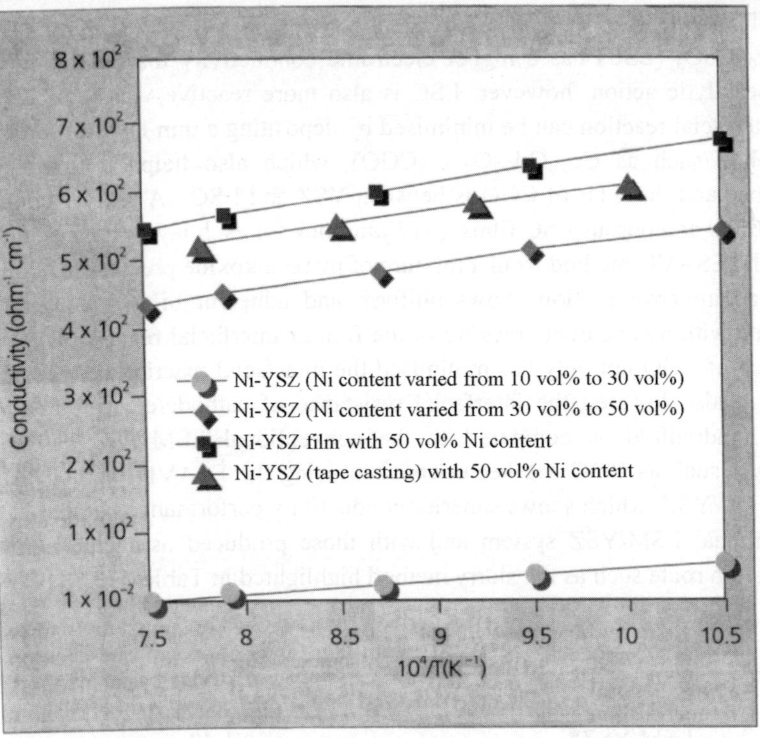

Fig. 14. Conductivity as a function of temperature for Ni-YSZ porous film with different Ni contents.

Uniform and porous compositionally-graded Ni-YSZ with varying nickel content have been deposited on to YSZ substrates using the ESAVD method by changing the Ni content in the solution gradually during the deposition process. Fig. 14 shows conductivity as a function of temperature for Ni-YSZ porous films with different Ni content. Although ESAVD can produce Ni-YSZ films (with 50 vol. % Ni) with higher conductivity than the conventional tape casting method, the compositionally graded Ni-YSZ (with Ni content varied between 30 vol. % and 50 vol.% at the outer edge of the anode film) also has remarkably high conductivity, and qualitative scratch testings show that the adhesion of the compositionally graded Ni-YSZ films is better than a single layer Ni-YSZ (with 50 vol.% Ni).

(e) Improved cathode/electrolyte systems

$La_{0.8}Sr_{0.2}CoO_3$ (LSC) has a higher electronic conductivity than LSE and superior electrocatalytic action, however, LSC is also more reactive with YSZ than LSM. The interfacial reaction can be minimised by depositing a thin layer of a doped ceria electrolyte such as $Ce_{0.8}Gd_{0.2}O_{0.19}$ (CGO), which also helps to lower the cell resistance and the CTE of CGO is between YSZ and LSC. A bilayer consisting of dense CGO and porous LSC films (~17 μm thick for each layer) has been deposited using the ESAVD method from a mixture of metal alkoxide precursors at 450°C [3]. The fracture cross-section shows uniform and adherent bilayer coatings on the substrate with no apparent cracking of the film or interfacial reaction layer [3]. The presence of CGO not only has minimised the interfacial reaction between LSC and YSZ but also lowered the interfacial resistance of cathode/electrolyte system by about hundredfold as compared to the conventional LSM/YSZ system. Other interlayer such as YDC has been deposited using the ESAVD method to produce LSC/YDC/YSZ which shows superior conductivity performance compared with the conventional LSM/YSZ system and with those produced using the conventional preparation route such as the slurry method highlighted in Table 1 [3].

Table 1: The electrode/electrolyte interfacial resistance of various systems deposited using ESAVD on to YSZ electrolyte substrates at 900°C.

System	R(900°C) Ω. cm²
LSM/YSZ*	1.48
LSC/YDC/YSZ	0.015
LSC/CGO/YSZ	0.012

YDC: $(Y_2O_3)_{0.15}(CeO_2)_{0.85}$

*Compared with the resistivity of LSM prepared by the conventional slurry method which was 4 Ω.cm² [36].

3.2. *ESAVD of thin films*

3.2.1. Oxide thin films

(a) Titanium dioxide

Anatase is one of the crystalline polymorphs of TiO_2, which has the potential for photoelectrical applications in semiconducting electrodes in solar cells, photocatalysis and semiconductor field-effect transistors, etc. The anatase thin films have been produced using sol-gel, CVD or PVD methods [37-39] and the thick films have been produced by conventional powder processing (e.g. screen printing or tape casting) [40] or colloidal (e.g. electrophoretic) [41] routes. These methods have the inherent limitations as discussed in Chapter 1. We have explored the use of ESAVD as a simple and cost-effective approach to synthesising anatase thin films. Titanium di-isopropoxide bis (2,4-pentanedionate) in 2-propanal has been used as the precursor in the synthesis of the films [42]. The formation of anatase films is strongly dependent on substrate temperature and precursor concentration. Fig. 15 shows the XRD traces of anatase films as a function of these process parameters. The films have been deposited on glass microscope slides, or SnO_2-coated glass substrates.

Anatase films can be produced at 400°C using 0.2M precursor solutions. At higher deposition temperatures (e.g. 450°C), the deposited anatase films tend to be opaque. However, uniform and optically transparent anatase films can be obtained at 450°C using a low precursor concentration (e.g. 0.05 M). The broad [101] peak in XRD pattern (Fig. 15) shows that the deposited films consist of nanocrystalline anatase TiO_2. The crystal size estimated from the Scherrer equation [43] is about 10nm and the Raman spectroscopy analysis confirms only anatase phase present in the deposited films. These films give good optical transmission over a wavelength of 370nm. However, at a lower deposition temperature (e.g. 400°C) and precursor concentration (e.g. 0.05 M), the deposited films are amorphous.

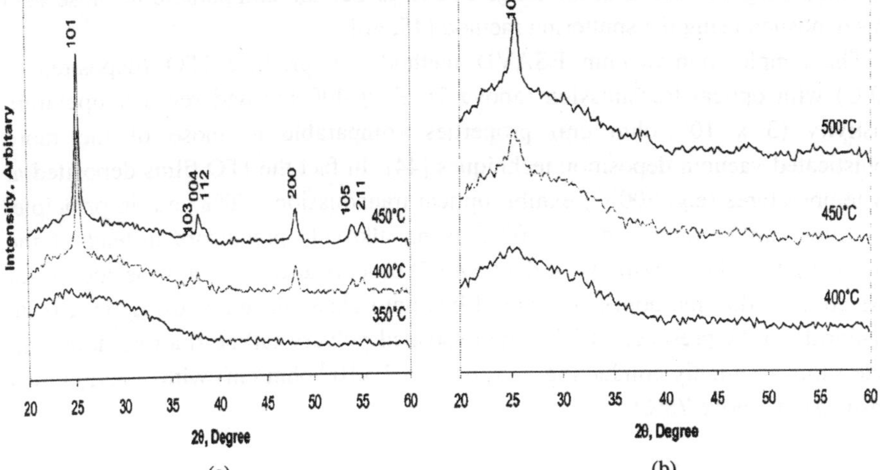

Fig. 15. XRD traces of TiO_2 films deposited at different temperatures (350-400°C) using (a) 0.2M and (b) 0.05M precursor solutions.

The flexibility of the ESAVD process also makes it possible to use additives such as polyethylene glycol (PEG) to control and modify grain size and surface morphology. Films with small amounts of low molecular weight PEG exhibit a structure of nano-sized TiO_2 particles, whereas a large amount (e.g 30wt%) of PEG can induce the formation of cubic anatase film with a large grain size of ~ 0.5 μm [42].

(b) Indium Tin Oxide

Indium Tin Oxide (ITO) thin films have potential applications in optical fields and optoelectronics including among other things low emission windows, rear view automobile windows and mirrors, antifrost coatings, solar cells, liquid crystal displays and photolumi nescence phosphor displays. A 0.05M mixture of tin and indium chlorides (with varying ratios, Sn:In =1:40 to 5 : 40) in methanol has been used as the precursor solution for the deposition of ITO [44] and ITO films have been deposited onto glass substrates. An increase in tin concentration increases the electrical resistivity of the films and the lowest resistivity has been achieved using the Sn: In =1 : 40. This ratio has been used to study the effects of temperature on ITO deposition. Films deposited at temperatures below 250°C are amorphous. Uniform and transparent crystalline ITO films have been deposited at 550°C and no further annealing is required. The XRD trace shows strong (222) and (400) peaks, which is in good agreement with other reported results [45,46]. For 0.5μm thick ITO film, the grain size is in the range of 200 to 600nm, comparable to those ITO films deposited using the sputtering method [45, 46].

The simple, non-vacuum ESAVD method can produce ITO (deposited at 550°C) with optical transmission (above 76%) at 400 nm and room temperature resistivity (3×10^{-4} ohm cm) properties comparable to those of the more sophisticated vacuum deposition techniques [44]. In fact the ITO films deposited at low temperatures (e.g. 300°C) exhibit optical transmission >70% and, in principle, ITO films can be produced at <300°C using alkoxide precursors instead of the chloride system. The flexibility of the ESAVD process also facilitates the successful fabrication of fluorine-doped tin oxide films onto glass substrates using tin acetate or chloride in the presence of HF. The deposited polycrystalline fluorine doped tin oxide films are highly conducting (7×10^{-3} to 2×10^{-4} ohm cm) with good optical transmission (above 78%).

3.2.2. Non-oxide thin films

We have synthesised non-oxide II-VI semiconducting thin films such as CdS and ZnS using ESAVD. Conventionally, these films have been produced using both vacuum deposition techniques such as CVD [47], PVD [48], and non-vacuum, low cost techniques such as spray pyrolysis [49], electrodeposition[50] and chemical bath deposition[51]. However, these conventional techniques have their limitations as described in Chapter 1.

(a) Cadmium sulphide

CdS thin films have been deposited by ESAVD from a mixture of cadmium chloride ($CdCl_2$) and thiourea [$(NH_2)_2CS$] precursors in an ethanol/water solvent [52]. Substrate temperatures below 500°C have been selected to prevent the oxidation of CdS, because oxygen is reported to have a detrimental effect on the electrical properties of the CdS films [53]. Dense and adherent CdS films have been successfully deposited onto glass substrates at low temperatures (≤450°C) in an open atmosphere and negligible oxygen and carbon contents have been found in these films which examined by X-ray photoelectron spectroscopy analysis. CdS films of high purity can thus be deposited by careful tailoring of the chemistry of the precursors and deposition conditions, without the need to use low pressure and/or a controlled atmosphere (e.g. inert gas) for the deposition environment.

The temperature and precursor composition have strong influences on the morphology and phases present in the films deposited by ESAVD. Such effects may be illustrated by CdS films [52, 54]. At optimum processing conditions, uniform CdS films with hexagonal phase and columnar structure and a preferred [002] orientation (see Fig. 16) were deposited at 400°C and with a growth rate of 0.05 μm/min. This was achieved using 0.005M precursor solution containing a Cd:S molar ratio of 1:3 at a flow rate of 10 ml/hr. Films deposited using a precursor solution containing a Cd:S molar ratio of 1:1, however, showed no preferred orientation as illustrated in Fig. 16(a).

XRD results show that the crystalline CdS film begins to form at a deposition temperature of 250°C and that the film mainly consists of hexagonal CdS and NH_4CdCl_3. Similar observations were reported by Krunks et al.[55]. The impurity phases such as NH_4CdCl_3 begin to disappear with increasing substrate temperature and the crystal size increases [54]. The crystal size of the film deposited at 300 °C is found to be 28nm and it increases to 70nm when the substrate temperature reaches 450°C. Fig. 17 shows the surface morphology of CdS films deposited at different temperatures (300-450°C) [52]. The films change from being porous with low crystallinity to being dense with high crystallinity with increasing substrate temperature. CdS films deposited at 300°C contain some pinholes and precipitates on the surface but the film became smoother with increasing substrate temperature. At the high temperature of 450°C however, some large particles formed on the smooth surface of the film.

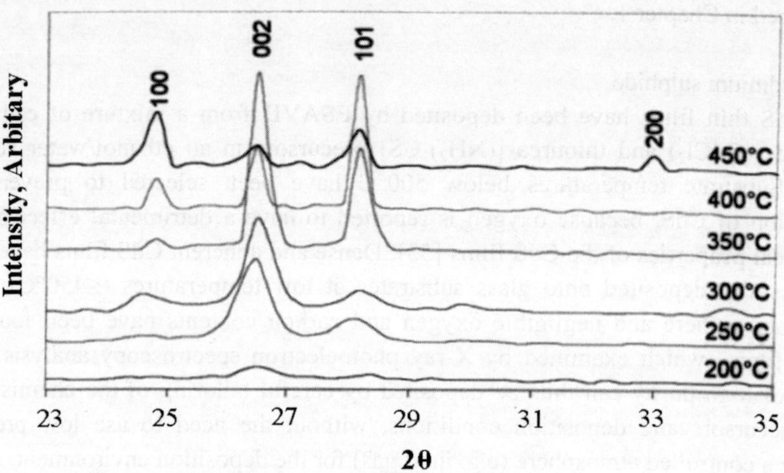

(a)

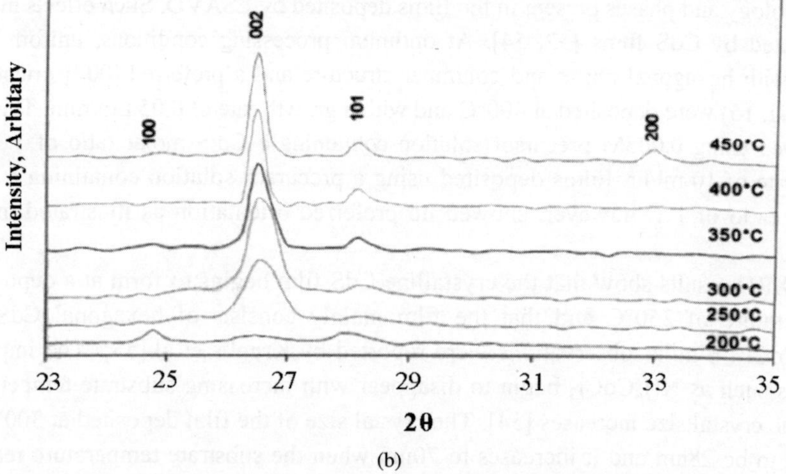

(b)

Fig. 16. XRD traces of CdS films deposited at different substrate temperatures (200-450°C) using different Cd:S ratios: (a) Cd:S = 1 : 1; and (b) Cd:S = 1 : 3.

The relationships between growth behaviour and film structure have been investigated, and a deposition model for the process has been proposed to explain the structural observations made [3, 54]. At low substrate temperatures (e.g. below 300°C), the charged aerosol is eletrostatically sprayed on to the substrate, and is followed by the removal of the solvent through evaporation and/or decomposition of the precursor. The resulting films tend to be porous and amorphous. At higher substrate temperatures (e.g. above 450°C), the solvent and chemical precursors vaporise and decompose before arrival at the substrate surface which results in the formation of particles in the gas phase. These particles are subsequently deposited onto the substrate surface and produce powdery films. At the intermediate temperatures (e.g. 300-450°C), both processes may occur depending on the substrate temperature. At an optimum substrate temperature, the solvent is evaporated near the heated substrate surface, and the precursor is subsequently vaporised in the vicinity of the substrate; it is then adsorbed on to the heated substrate surface and decomposition and/or chemical reactions occur to yield the desirable dense and columnar CdS film with a good adhesion to the substrate, resulting in a typical heterogeneous CVD reaction (as shown in Zone D in Fig.1). Obviously, the optimum temperatures are found in a range between 400 and 450°C for ESAVD of CdS films. The columnar structure is desirable to enhance electrical conductivity because of the reducing effects of grain boundaries on electron mobility across the films [56].

The optical properties of the films are measured by a SHIMADZU UV-1601 spectrophotometer covering a spectral range from 200 to 1000nm. The electrical properties of the films are characterised using Van der Pauw's method. The optical properties of the CdS films produced by ESAVD are found to be very similar to those produced by other vacuum deposition methods [52]. The direct band-gap value is 2.43~2.44eV, which is in agreement with the reported value for single-crystal CdS [57]. The Cd:S molar ratio is found to have very little influence on the optical properties but has a tremendous effect on the electrical conductivity of the CdS films. The electrical conductivity of CdS films ranges from 2.5×10^{-3} to 1.7×10^{-5} $(\Omega\ \text{cm})^{-1}$, and increases with increasing Cd:S molar ratio. This may be due to the better crystalline structure of the films as a result of an excess of thiourea in the precursor solution. Excess thiourea is needed to produce the $Cd(SCN_2H_4)_2Cl_2$ intermediate complex which facilitates a complete conversion to form fully crystalline CdS films with less stacking faults [52]. The results have shown that the

ESAVD technique can produce CdS films with optical and electrical properties suitable for solar cell applications.

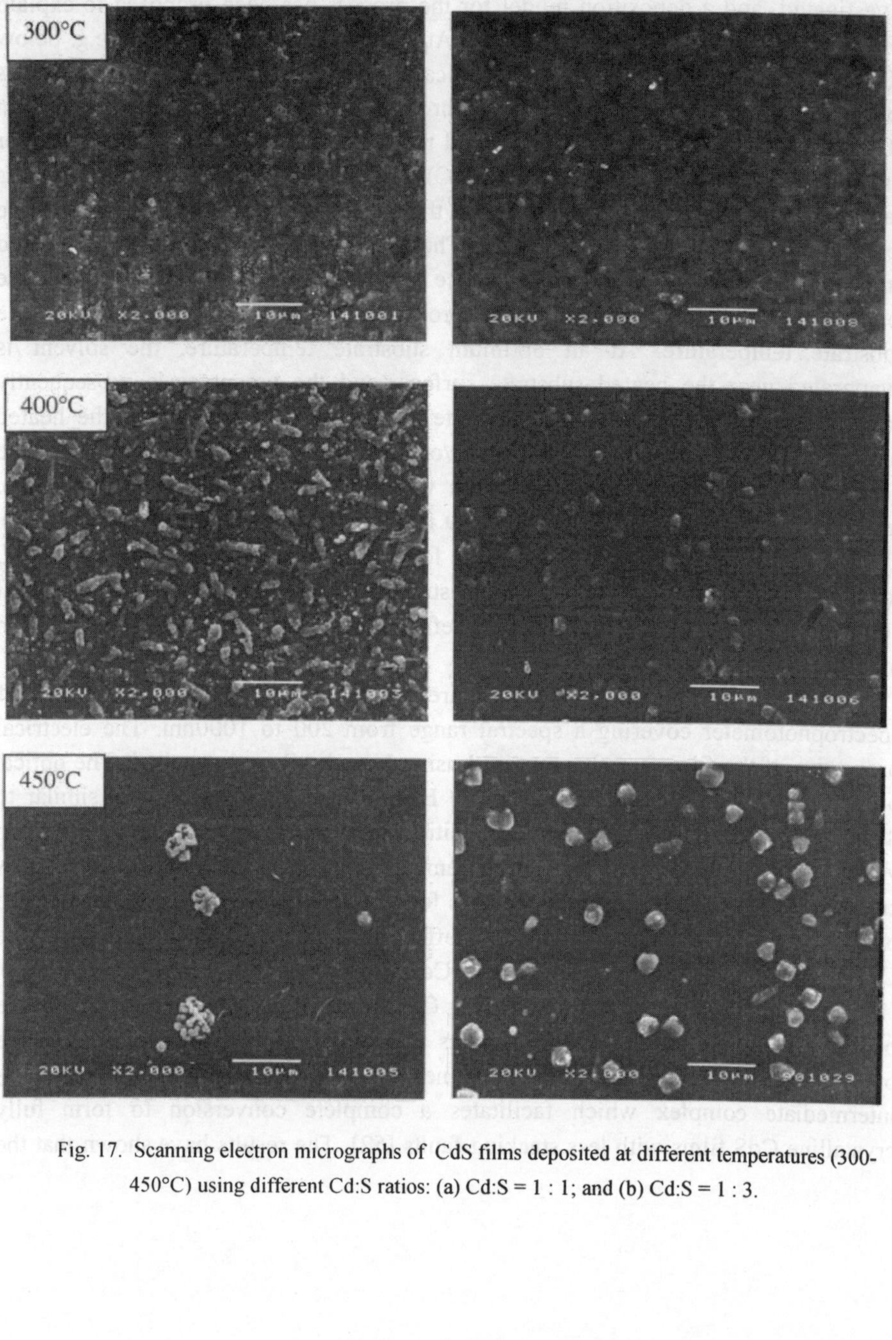

Fig. 17. Scanning electron micrographs of CdS films deposited at different temperatures (300-450°C) using different Cd:S ratios: (a) Cd:S = 1 : 1; and (b) Cd:S = 1 : 3.

(b) Zinc sulphide

In addition to the polycrystalline growth, ESAVD is capable of depositing highly orientated thin films and epitaxial films. The wide band gap (~3.8eV) and good luminescence characteristics of ZnS have made it attractive for optoelectronics applications such as light emitting diodes and flat panel electro-luminescent displays. Vacuum deposition techniques such as molecular beam epitaxy and pulsed laser deposition are commonly used to deposit highly textured ZnS films on to single crystal substrates [58, 59, 60].

We have demonstrated the viability of depositing highly textured ZnS thin films onto single crystal as well as amorphous substrates using the non-vacuum ESAVD method. The textured ZnS thin films have been synthesised using a mixture of 0.01-0.05 M of stoichiometric $ZnCl_2$ and $(NH_2)CS$ precursor solutions, and depositing at temperatures between 450-500°C. Both amorphous [e.g. glass] and single crystal [e.g. (100) Si] substrates have been used in the investigation.

(i) Amorphous substrate

Fig. 18 shows the XRD traces of ZnS films deposited at 450 and 550°C. At a lower deposition temperature of 450°C, the deposited ZnS films have a very small nanocrystalline grain size (<20nm), determined by the Scherrer equation. Films deposited at a higher temperature of 500°C exhibit clear evidence of the textured nature of the ZnS films. This is further supported by the pole figure analysis where the [111] is normal to the surface of the sample from 111 pole figure analysis. The selected area diffraction pattern from the TEM analysis (see Fig.19) also confirms the [111] textured nature of the film and shows the films containing the cubic phase [61].

(ii) Single crystal substrate

Highly orientated ZnS films have also been grown on (100) Si substrates. ZnS has a reasonably good lattice match (0.2% match) with the single crystal substrate and the as-deposited films exhibit a strong cubic (200) peak as shown in the XRD trace (Fig. 20). This shows that ZnS has a preferred [100] growth orientation to the (100) Si substrate compared with the preferred [111] orientation on the amorphous glass substrate and there is therefore a tendency towards epitaxial crystal growth in ZnS where the template crystalline structure is commensurate with the nucleating crystal face, in this case an epitaxial relationship of $[100]_{Si}//[100]_{ZnS}$. This epitaxial growth tendency has also been confirmed by TEM and HRTEM studies [62].

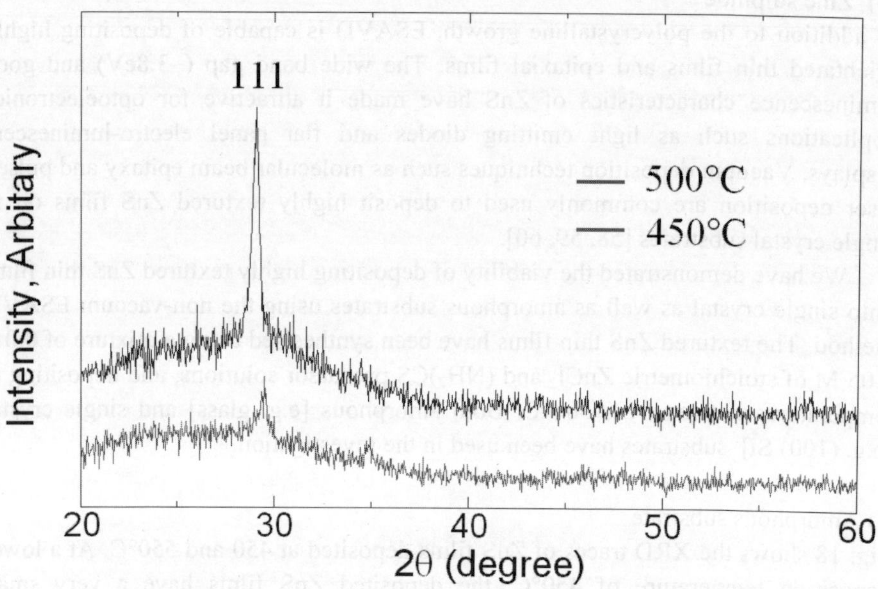

Fig. 18. XRD traces of ZnS films deposited at 450°C and 500°C.

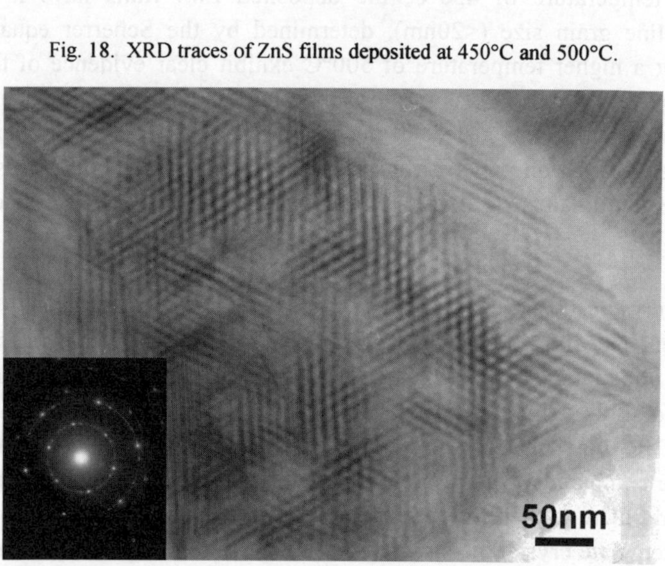

Fig. 19. A TEM image showing a plan of a ZnS film on a glass substrate.

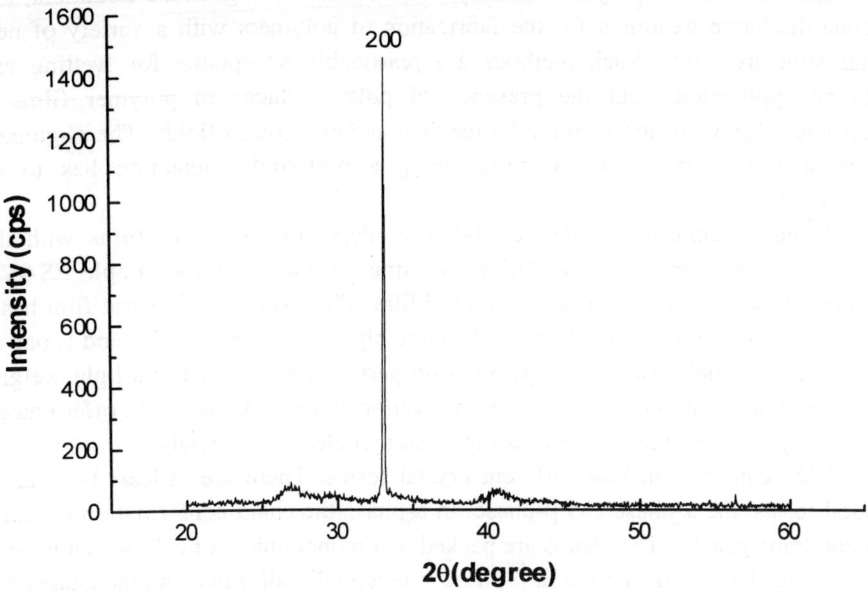

Fig. 20. XRD pattern of the ESAVD of a ZnS film on a (100) Si substrate.

This unique texture and/or epitaxial formation may be due to the alignment of dipolar units induced by the electric field during the decomposition and restructuring of the precursors in the deposition process which results in the development of a preferred orientation or texturing in the films.

(c) Polyvinylidene fluoride (PVDF)

Such preferred orientation in growth has also been observed in the ESAVD of polymeric films with specific surface structure and chemical composition. The control of the surface properties of the polymer is essential to the service performance of engineering devices such as printing, wetting, biomedical and electronic instruments. This is because molecules from the 'environment' approach the polymer surface and experience interfacial forces arising from the interaction between the electrostatic and positively- or negatively-charged cloud. It is thus possible in principle to design 'guided' approaches to device optimisation by modifying the chemical structure at the surface layers to provide controllable interfacial interaction between polymers. The conventional surface modification techniques used in the polymer industry include acid etching, flame treatment, and corona discharge treatment for the fabrication of polymers with a variety of new polar structures [63]. Such methods are reasonably acceptable for wetting and printing applications, but the presence of polar surfaces in polymer films is insufficient for applications in the biomedical and electronics fields. The alignment of polar groups in polymeric films along a preferred orientation has to be considered.

We have demonstrated the feasibility of depositing polymer films with the required crystal forms and orientation by using the modified and simple ESAVD method, which is demonstrated in a PVDF film. This type of polymeric film has a large dielectric constant, arising from the large dipolar moment of CF_2, and is one of the most polar dielectric polymers. Such properties, together with its light weight, good flexibility and toughness, ease of fabrication and low permittivity, offer unique advantages over ceramics as piezoelectric and pyroelectric materials.

PVDF can exist in four different crystal forms. There are at least two stable crystal forms, the α-phase and β-phase. In α-phase the chain conformation is trans-gauche-trans-gauche. The chains are packed in a monoclinic unit cell, which is non-polar. The chain confirmation in β-phase is essentially all trans, and the chains are crystallised in an orthorhombic unit cell with a net dipole moment. The polar β-phase is essential for good piezoelectric and pyroelectric properties in PVDF films [64, 65]. The conventional fabrication of PVDF films for electronic applications normally involves several production steps: first PVDF bulk films are produced by a conventional method such as casting, hot pressing, dipping and spin coating of the PVDF solution and then the PVDF bulk films are monoaxially or biaxially stretched to produce the β-phase and finally poled under high voltage, corona or electron beam discharge. Since PVDF films already deposited by dipping or spin coating from solution cannot easily be stretched, it is difficult to prepare β-phase crystal films by mechanical means, however, a certain amount of β-phase content can be produced by direct corona discharge or high electrical field exposure [66a].

The ESAVD process has been adapted for the deposition of a highly orientated β-phase content using corona poling in a single production step: 0.01-0.1M PVDF solution (PVDF powder dissolved into N,N,-dimethylformamide or N,N,-dimethylacetamide solvent) was corona-sprayed towards a Si or metallised Si substrate and heated at about 150°C; the vaporisation of the solvent in the vicinity or on the substrate surface occurs in the presence of the corona field during the deposition. The precursor flow rate was varied between 0.4 and 15ml h^{-1}. By maintaining the corona field (Process A) during the cooling of the coated sample after the deposition, thin films with high β-crystallite content (at 2θ ~20.8°) have been produced compared with those where the corona field has been switched off (Process B) during sample cooling, as shown in the XRD traces (Fig. 21). Such a result is also supported by the infrared spectroscopy analysis which shows a high β-phase content (at ~530 cm^{-1} and ~1280 cm^{-1}) in PVDF films produced by Process A [66b]. The IR results also suggest that the CF_2 dipoles (~1280 cm^{-1}) are aligned along the applied corona field.

The desired polar β-form crystal exhibits a net dipole moment and it is thus reasonable to propose that molecular dipoles on chains within the crystalline regions of the polymer become aligned with the applied electric field during the corona poling process. Such utilisation of the corona field during the deposition process and cooling down stage is effective in controlling crystal forms and their orientation.

The unique capacity of ESAVD for fabrication of polymer films with specific surface structure and chemical composition for optimum performance allows the production of polymers with tailored surface compositions for particular applications which are currently sought in the polymer industry.

3.3. Other variants of ESAVD

3.3.1. Electrostatic Assisted Aerosol Jet Deposition (EAAJD)

Electrostatic Assisted Aerosol Jet Deposition (EAAJD) is a variant of the ESAVD process [25]. Fig. 22 shows a schematic diagram of the EAAJD process in which the precursor solution is atomised into aerosol droplets with fine particle size <2μm, using ultrasonic excitation (1.7 MHz). A carrier gas such as nitrogen (flow rate about 0.5 to 2 l/min) is used to facilitate the transport of aerosol into a corona discharge unit to form a charged aerosol which is subsequently delivered towards a heated substrate under an electric field (e.g. 2 –5 kV/cm), where it undergoes decomposition and/or heterogeneous chemical reactions near or on the substrate surface to deposit a dense and adherent film. For the synthesis of ultrafine powders, the charged aerosol is delivered to a heated zone instead in order to tailor the homogeneous gas phase reaction. The EAAJD process thus has a similar deposition mechanism to that of the ESAVD method, and it can be used to fabricate both films and powders; the main difference between the EAAJD and the ESAVD methods however is that the aerosol atomisation and electrostatic discharge functions in EAAJD are separated, whereas in the ESAVD process they are not. This allows the use of a wider choice of precursors (both aqueous and non aqueous precursor systems), as well providing a higher deposition efficiency (>95%), and deposition rate (> 2 μm/min), along with the advantages of ESAVD as outlined in section 2.8.

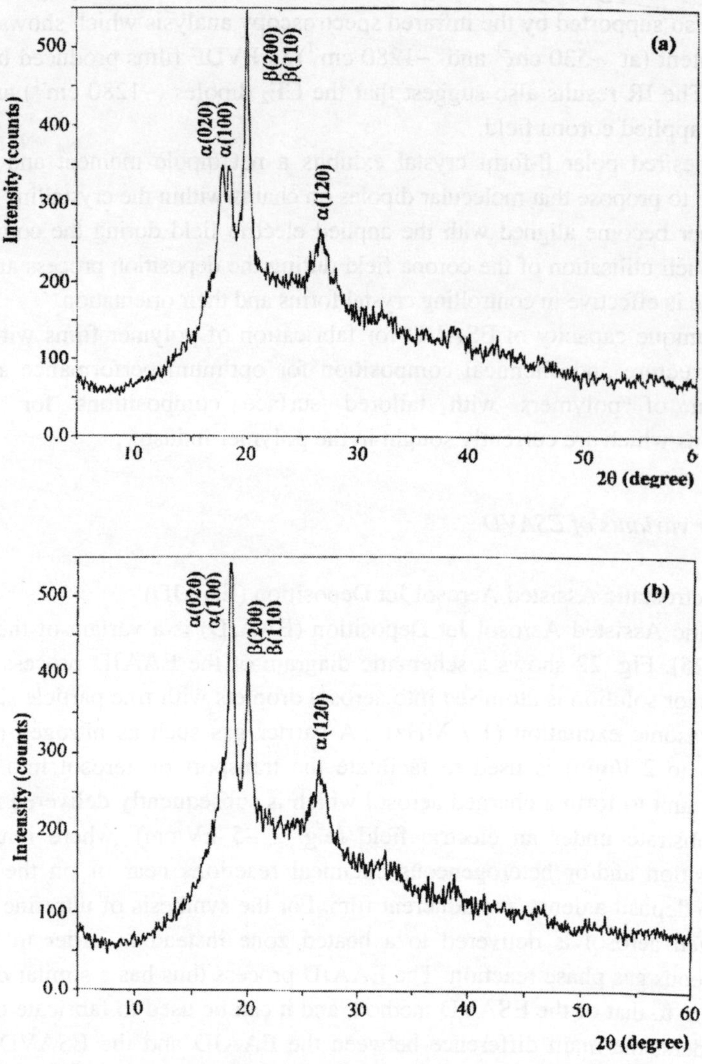

Fig. 21. XRD traces of the PVDF film deposited by the ESAVD method using (a) Process A, and (b) Process B.

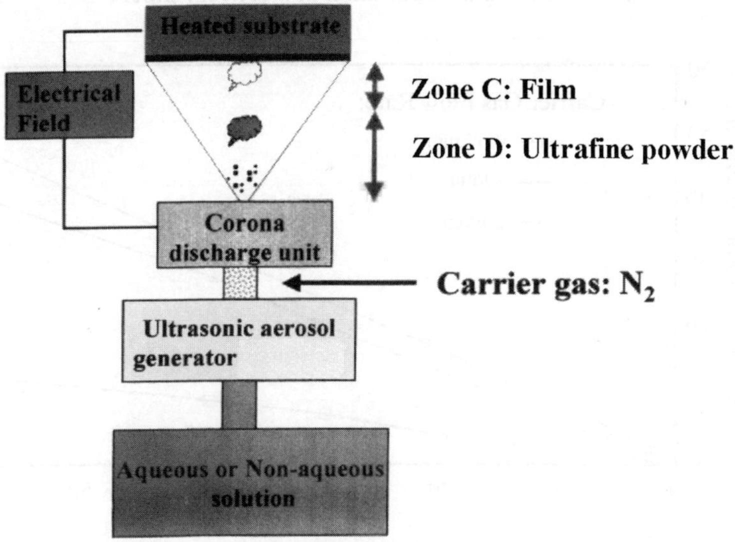

Fig. 22. A schematic diagram of the EAAJD process.

A narrow distribution of aerosol droplets is desirable to faciliate decomposition and chemical reaction simultaneously, to enhance coating uniformity or the formation of uniform particles. The use of ultrasonic atomisation allows a high generation rate of fine droplets with a narrow droplet size distribution compared with pneumatic spraying. The droplet size can be reduced by increasing the exicitation frequency, f, of the ultrasonic atomiser, as shown in the equation established by Lang: $d=k(8\pi\sigma/\rho f^2)^{1/3}$ [67], where k is a constant, σ is the surface tension and ρ the density. The rate of nebulisation depends on the saturation of vapour pressure, P, the viscosity coefficient, η, and coefficient of surface tension, τ; the proportionality factor, β, is given by the equation $\beta = P/\eta\tau$ [68]. The aqueous solvent (e.g. H_2O) has a higher rate of aerosol generation than the organic solvent (~1 g/min) and does not contain carbon, rendering the incorporation of carbon into the films unnecessary in films deposited at a lower temperature. It is also cheaper, safer and more environmentally friendly than organic solvents. Experimentally it has been found that the volume of aerosol generated using the aqueous solution increases linearly with time, and with the gas flow rate as shown in Fig. 23 [69]. The use of a higher gas flow rate also helps to minimise the condensation of aerosol during transportation in a delivery pipe.

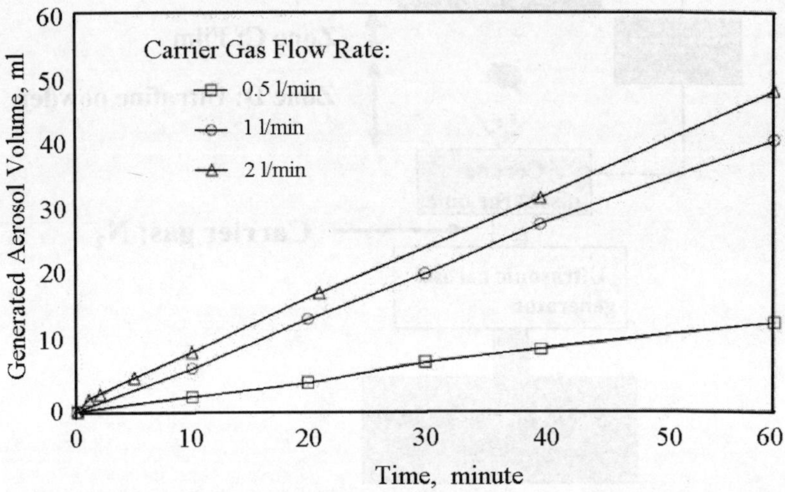

Fig. 23. The rate of aerosol generation for the aqueous solution using different carrier gas flow rates.

The use of the EAAJD method to synthesis II-VI semiconducting thin films and powders has been reported [25]. Non-oxide II-VI semiconducting films such as CdS, CdSe and ZnS have potential applications in devices based on electroluminescent and photoluminescent phenomena [70], and sharp cut filters in the visible and ultraviolet regions [71]. These films have been deposited using EAAJD in an open atmosphere by carefully tailoring the chemistry of the precursors and so it avoids the use of a reduced-pressure processing environment. This can help to lower the cost of fabrication while maintaining the quality of the films. Fig. 24 shows the XRD patterns of these films which have been deposited at low temperatures using a mixture of metal chloride and thiourea solutions (0.01 M) for CdS and ZnS, and a mixture of metal chloride and selenourea solutions (0.01 M) for CdSe [69]. Both CdS and ZnS films exhibit strong preferred orientations, with c-axis normal to the film surface. TEM results indicate that the CdS film deposited by EAAJD is nanocrystalline and exhibits a hexagonal structure with a strong preferred (002) orientation, independent of the Cd/S molar ratio and substrate temperature. This result is similar to those CdS films synthesised using the ESAVD method. AFM images in Fig. 25 show that the grain size for the CdS and CdSe films is ca. 50-200 nm, while that for ZnS film is larger (ca. 300-500) [69]. The optical properties for the films show that their bandgap energy values are 1.65, 2.42 and 3.30 eV respectively for the CdSe, CdS and ZnS films.

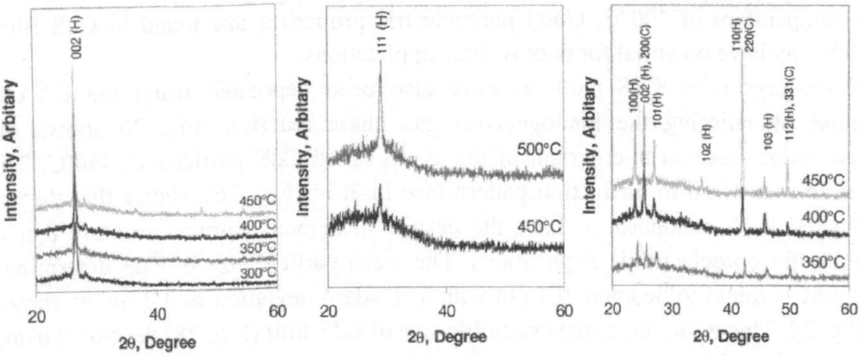

Fig. 24. XRD traces of (a) CdS, (b) ZnS, and (c) CdSe films, deposited at different substrate temperatures.

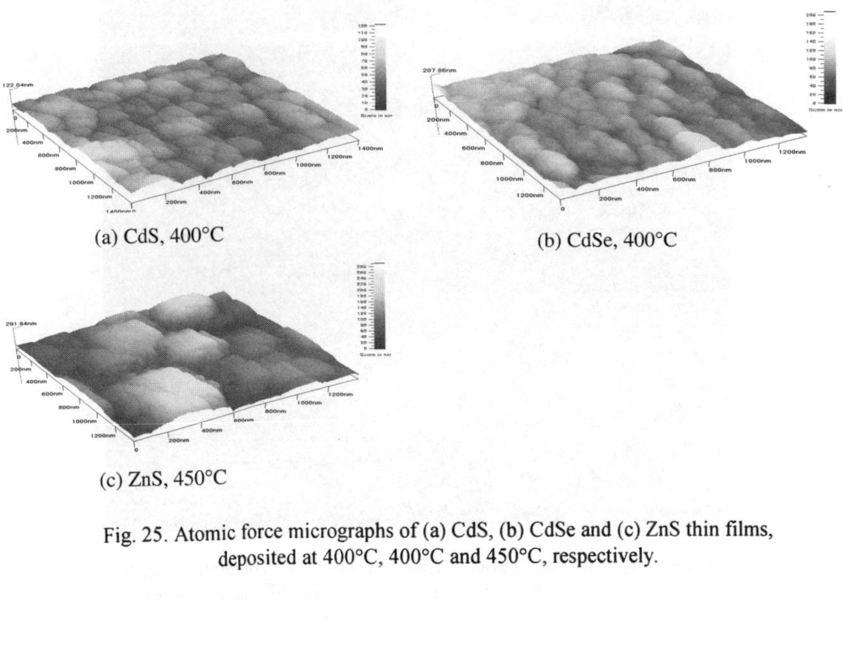

(a) CdS, 400°C

(b) CdSe, 400°C

(c) ZnS, 450°C

Fig. 25. Atomic force micrographs of (a) CdS, (b) CdSe and (c) ZnS thin films, deposited at 400°C, 400°C and 450°C, respectively.

The bandgap values of CdS films deposited at 300, 350, 400 and 450°C are found to be 2.44, 2.42, 2.42, 2.41 eV, respectively [72]. The blue shift ΔE_g (0.02 eV) from the bulk value (2.42 eV) is believed to be due the quantum size effect resulting from the reduction in crystallite size (< 8 nm) in CdS films deposited at a low temperature of 300°C. Good photoelectric properties are found in CdS films which may have potential for photovoltaic applications.

Nanocrystalline CdS particles have also been deposited using the EAAJD method by tailoring the homogeneous gas phase reaction. Fig. 26 shows the transmission electron micrograph of the as-deposited CdS particles at 450°C. The selected area electron diffraction pattern (see inset in Fig. 26), shows that the as-deposited II-VI nanoparticles have the desired hexagonal wurzite structure that is suitable for optoelectronic applications. The mean particle size of CdS determined by TEM is found to be about 9.9 nm with a standard deviation of 3.2 nm as shown in Fig. 27. This is similar to the crystallite size of CdS film (Fig. 28) deposited using a similar deposition temperature but tailoring the heterogeneous chemical reactions instead.

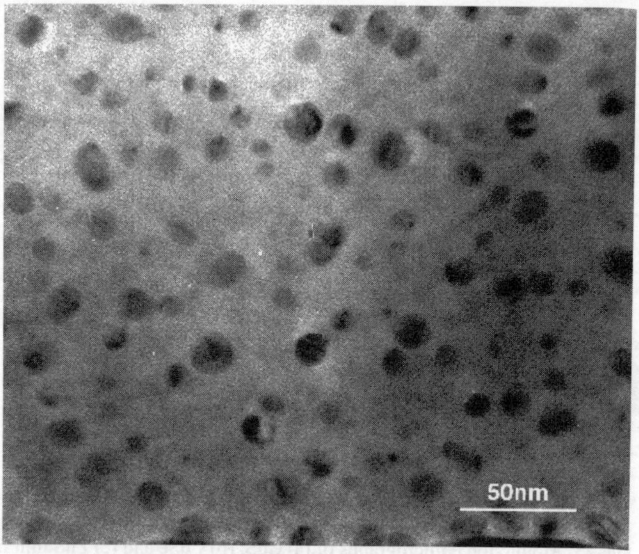

Fig. 26. A TEM image of CdS nano-particles deposited at 450°C. The inset shows the selected area diffraction pattern resulting from the hexagonal CdS nanoparticles.

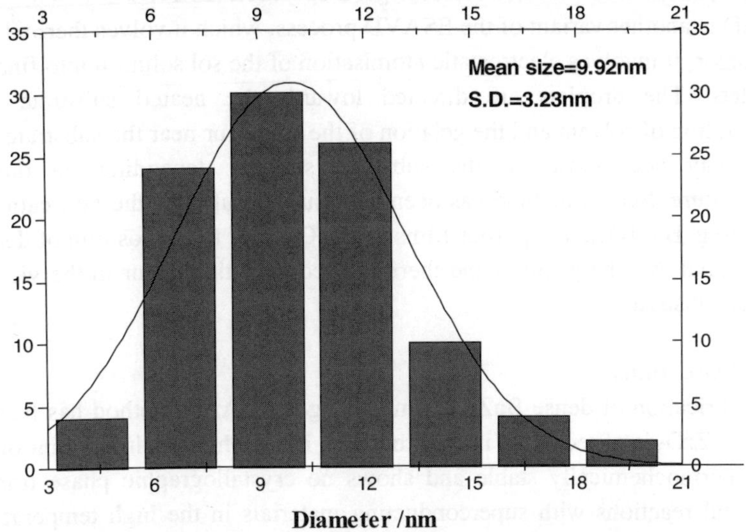

Fig. 27. Particle distribution of CdS nanopowder, synthesised at 450°C.

Fig. 28. A TEM image plan of CdS thin film deposited at 450°C. The inset shows the selected area electron diffraction pattern of the film.

3.3.2 Electrostatic Assisted Aerosol-gel Deposition (EAAD)

EAAD is another variant of the ESAVD process, which involves thermal gelation of precursor. It involves electrostatic atomisation of the sol solution into finely charged droplets. The droplets are directed towards the heated substrate where the evaporation of solvent and the gelation of the sol occur near the substrate surface, or they may decompose on the substrate surface, depending on the substrate temperature. Such a method has been used successfully for the fabrication of dense films (e.g. $BaZrO_3$), and porous films (e.g.SiO_2). For the deposition of dense films it is essential that the gelation and thermal decomposition occur in the vicinity of the heated substrate.

(a) Dense films

The fabrication of dense $BaZrO_3$ films using the EAAD method has been reported [73]. $BaZrO_3$ is of technical interest because it has a high melting point of ~2600°C. It is thermochemically stable and shows no crystallographic phase transitions or chemical reactions with superconducting materials in the high temperature region found in high Tc superconductor processing. It is thus ideal container material for the solution growth of single crystals of $YBa_2Cu_3O_{7-\delta}$ (YBCO), a substrate or buffer layer for YBCO superconductors, and for diffusion barriers in Bi(2223) multifilament tapes to reduce ac losses caused by interfilamentary coupling. $BaZrO_3$ has poor sinterability and materials made from sintered $BaZrO_3$ ceramics with high densities are very brittle and fragile, and prone to crack. The manufacture of high density bulk $BaZrO_3$ material is thus very difficult and expensive [74] and applying $BaZrO_3$ coatings or films to substrates is another approach to address this fabrication problem. Sol-gel and hydrothermal synthesis methods have been used to fabricate $BaZrO_3$ thin films but the thickness of the films has been limited to about 100-300 nm [75, 76]. Thick $BaZrO_3$ films have been fabricated using reactive liquid phase sintering, plasma spraying [77] and polymer painting [78] but none of these methods has been successful and the properties of the films require further improvement.

We have demonstrated the usefulness of the EAAD method in fabricating dense and thick $BaZrO_3$ film. The precursor used is synthesised by reacting barium with 2-methoxyethanol and then mixing with zirconium n-propoxide in a 2-methoxyethanol solvent. The chemical reactions are as follows [73]:

$$\begin{aligned}&Ba(s) + 2MeOCH_2CH_2OH(l) \rightarrow \\ &Ba(OCH_2CH_2OMe)_2(l) + H_2(g)\end{aligned} \tag{5}$$

$$Zr(OPr^n)_4(l) + 4MeOCH_2CH_2OH(l) \rightarrow$$
$$Zr(OCH_2CH_2OMe)_4(l) + 4Pr^nOH(l) \quad (6)$$

The precursor is electrostatically atomised to form the aerosol which is an oxo-alkoxide complex and comprises a BaZrO core surrounded by several $(OR)_m$ and $(ROH)_n$ ligands and carboxylate ions CO_3^{2-} arising from exposure to air. Although single phase $BaZrO_3$ films are formed at temperatures above 800°C, these films are porous with poor coating adhesion. Films with large cracks have been deposited at low substrate temperatures (<380°C). The optimum temperature is found to be about 400°C where aerosol gelation and decomposition occur simultaneously in the vicinity of the heated substrate. The as-deposited films are crack-free and mainly amorphous with a small amount of microcrystalline $BaCO_3$. Further heat treatment is required to produce single-phase $BaZrO_3$ films. Repeated deposition and firing (three cycles) was required to produce thick (20 μm), crack-free and crystalline $BaZrO_3$ films as shown in Fig. 29.

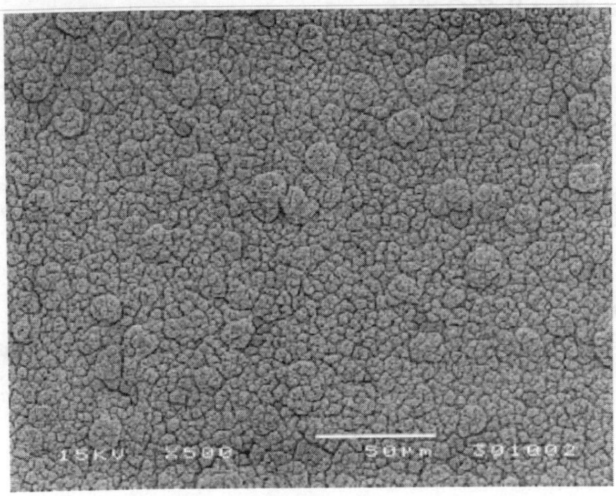

Fig. 29. The surface morphology of a $BaZrO_3$ film. The film thickness is ca. 20 μm.

The structural evolution has been investigated using a combination of TG-DTA, FTIR and XRD. The morphological development and deposition mechanism are strongly influenced by the substrate temperature, which is illustrated by a schematic diagram, Fig. 30.

T < 200°C: At low substrate temperatures, the precursor (i.e. oxo-alkoxide complex) is thermally decomposed to form the ROH ligands and deposited on to the substrate. This temperature is insufficiently high to vaporise the solvent during the transportation of the droplets towards the substrate and as a result a precipitate is formed after solvent evaporation as shown in Fig. 30(a).

200°C < T < 400°C: The increase in substrate temperature causes solvent evaporation and thermal gelation to occur during the transport of the aerosol towards the heated substrate, and a wet film is formed with large cracks owing to rapid shrinkage during aerosol-gel transformation Fig. 30(b). This film consists of amorphous $BaZrO_3$ as well as the intermediate $BaCO_3$ phase which is due to the decomposition of OR ligands at temperatures above 350°C.

T = 400°C: At this temperature, the solvent is almost vaporised before arriving at the substrate surface. The aerosol gelation and decomposition occur simultaneously in the vicinity of the heated substrate leading to the deposition of a crack-free film (Fig. 30(c)). The as-deposited film is amorphous with a small amount of microcrystalline $BaCO_3$. Further heat treatment at 800°C is required to produce a single-phase $BaZrO_3$ film.

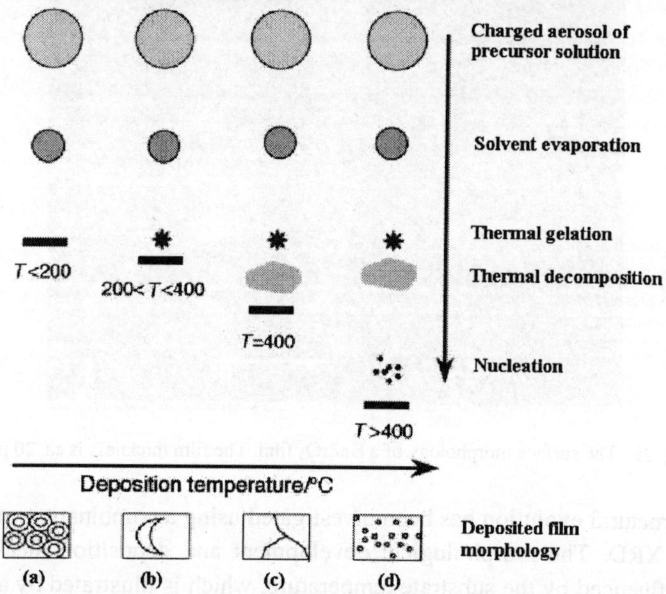

Fig. 30. A schematic diagram of the mechanism of film formation during the EAAD process and morphology as a function of deposition temperature.

T > 400°C: A single phase $BaZrO_3$ film is formed at substrate temperatures above 800°C. However, at a high substrate temperature (above 400°C), solvent evaporation, thermal gelation and thermal decomposition occur in the gas phase before arrival at the substrate surface, resulting in the formation of solid particles in the gas phase which subsequently deposit onto the substrate, leading to the fabrication of porous films with poor coating adhesion [Fig. 30(d)].

(b) Porous films

Porous silica films can be used as catalysts supports, sensors, filters and host matrices for semiconductor and high speed integrated circuits, packaging etc. These films are generally produced using sol-gel processing from a silicon alkoxide precursor (e.g. tetraethylorthosilicate, TEOS) under acidic or basic conditions [79]. Other methods such as spin coating and electrophoretic deposition using colloidal sol have also been used to produce thick silica films [80, 81], but these conventional methods have limitations as discussed in section Chapter 1.

We have explored the feasibility of producing porous silica films using the Electrostatic Assisted Aerosol-gel Deposition (EAAD) method. A commercial silica sol (Ludox AM 30, Dupont USA) with a 30wt% SiO_2 content in an aqueous solution has been used as the starting precursor. The sol consists of primary particle size (13nm) with a monodistribution; the sol particle surface is negatively charged, with Na^+ as a counter ion to stabilise the colloidal solution. The commercial sol solution is diluted with ethanol at a SiO_2 content of 0.2-1.0 g l^{-1}. A silica gel film is deposited at a substrate temperature between 80 and 120°C, followed by subsequent heat treatment at 500°C for 1h to produce the crack-free, porous silica film with a three-dimensional network of spherical pore structure.

The pore size and inter-pore wall thickness increases with increasing flow rate. For example, the increase of precursor flow rate from 5 ml h^{-1} to 10 ml h^{-1}, leads to an increase in pore diameter from 5μm to 15μm, and the inter-pore wall thickness increases from 2μm to 4μm (see Fig. 31). Such observation is consistent with the Scaling law in electrospraying [11], where droplet size increases with increasing flow rate which results in the formation of a larger pore structure in the film. The FTIR spectra shows that the silica films produced from the commercial sol solution consist of a Si-O-Si network with very little non-bridging Si-O and Si-OH groups [82] and the silica films synthesised by EAAD from the commercial silica sol solution are therefore hydrophilic. This property is desirable for applications such as sensor matrices [83] and high speed integrated circuit packaging [80]. Whereas, the films derived from Si alkoxide precursor such as TEOS are hydrophilic [84] which results in an unacceptably high dielectric loss (loss tangent >0.5) or high moisture absorption.

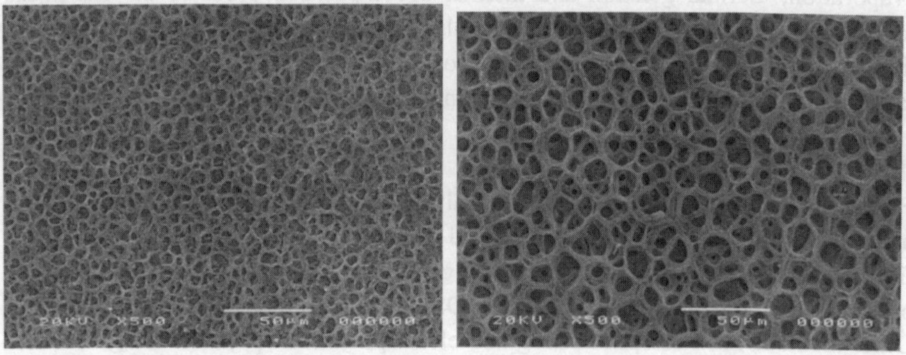

(a) (b)

Fig. 31. Scanning electron micrograph of porous silica films with reticular structures. The films have been deposited at 100°C with a silica content of 0.2 g l^{-1} at different flow rates: (a) 5 ml h^{-1}, and (b) 10 ml h^{-1}.

Deposition mechanism

The deposition mechanism of porous silica films using the EAAD method has been proposed and is illustrated with a physical model as shown in Fig 32 [82].

When the colloidal silica solution consisting of silica sol particles with negatively charged surfaces (stabilised by Na^+ as a counter ion) passes through a positively charged nozzle, it is atomised electrostatically into micron and sub-micron-sized droplets and the negatively charged sol particles are attracted to the surface of positively charged droplets. Any solvent evaporation or gelation of the droplets is insignificant during the transportation from the nozzle to the heated substrate as the substrate temperature is relatively low (<120°C). As the droplets hit the substrate, they discharge and the heat transfer from the substrate to the droplets results in the evaporation of the solvent and the gelation of the sol to form a reticular structure.

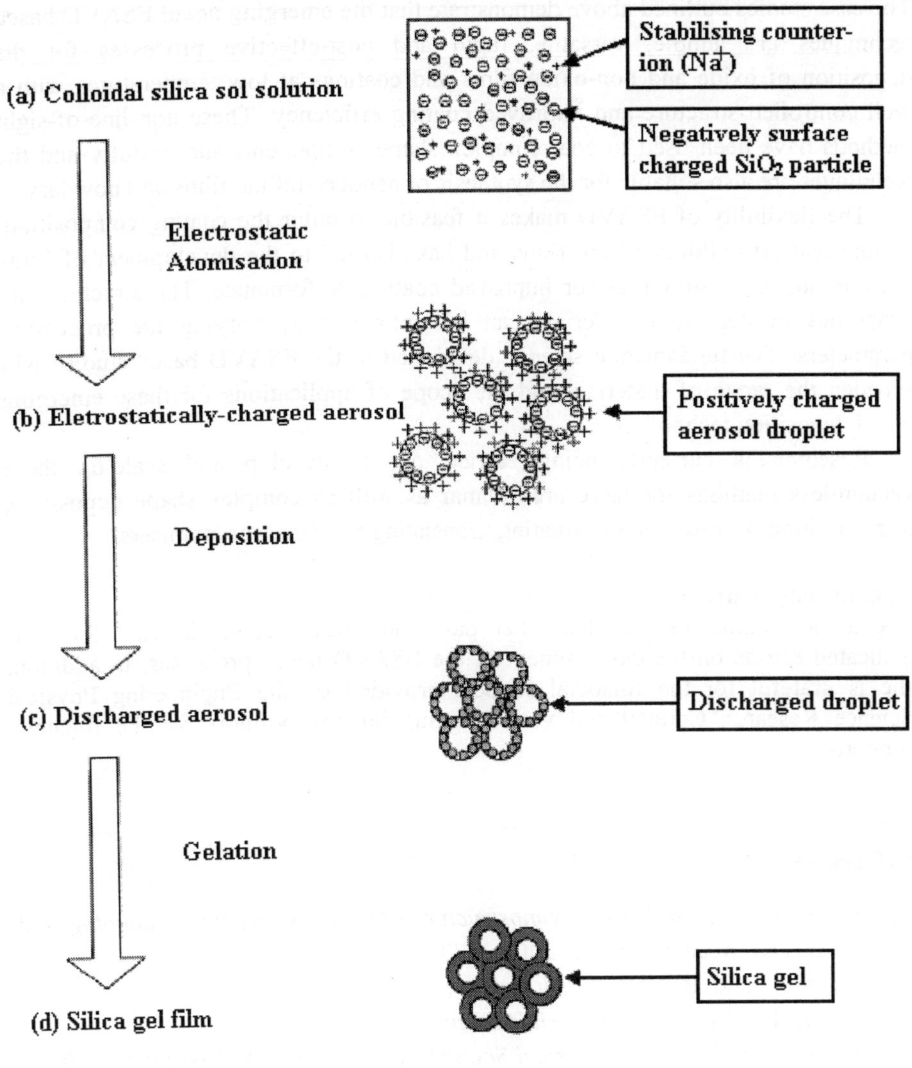

Fig. 32. A schematic diagram of the proposed deposition mechanism of porous silica films synthesised by the EAAD method.

4. Conclusions and Outlook

The case studies outlined above demonstrate that the emerging novel ESAVD based techniques are simple, versatile, rapid and cost-effective processes for the deposition of oxide and non-oxide films and coatings at low temperatures with a well controlled structure and improved coating efficiency. These non-line-of-sight methods have been used to coat complex shape components successfully and the techniques are also suitable for the synthesis of nanocrystalline films and powders.

The flexibility of ESAVD makes it feasible to tailor the coating composition, texture and crystallinity of the films and has also led to the development of films with unique microstructures for improved coating performance. The structure and properties of deposited materials can be optimised by varying the processing parameters. The fundamental science developed in the ESAVD based studies will broaden the range of materials and the scope of applications for these emerging novel deposition technologies.

Research is currently being carried out to develop and scale-up these vacuumless methods for large area planar as well as complex shape deposition, using multiple atomiser and/or rotating, translating substrates or atomisers.

Acknowledgments

The author would like to thank her past and present research team for their dedicated efforts on the development of the ESAVD based processes. In addition, she is grateful for the financial support provided by the Engineering Physical Science Research Council and various industrial collaborators for the financial supports.

References

1. Choy K. L., *in Handbook of Nanostructured Materials and Nanotechnology*, ed. Nalwa H.S. (Academic Press, 2000), 533.
2. Choy K. L. and Bai W., *British Patent* (1995), 9525505.5.
3. Choy K. L., *Progress of Materials Science*, (2001, in press).
4. Zeleny J., *The American Physical Society, Second Series*, **Vol III** (1914), 69.
5. Grace J. M. and Marijnissen J. C., *J. Aerosol Sci.* **25** (1994), 1005.
6. Cloupeau M. and Prunet-Foch B., *J. Electrostatics* **22** (1989), 135.
7. Taylor G. I., *Proc R Soc London Ser* A **280** (1964), 383.
8. Fernández de la Mora J and Loscertales I. G., *J Fluid Mech* **260** (1994), 155.
9. Gañán-Calvo A. M., Dávila J. and Barrero A., *J Aerosol Sci* **28** (1997), 249.

10. Choy K. L. and Su B., *Thin Solid Films* **388** (2001), 9-14.
11. Ganan-Calvo A. M., *J. Aerosol Sci.* **25 Suppl. 1** (1994), S309.
12. Grigoriev D. and Choy K. L., *Phil. Mag* (submitted).
13. Choy K. L., *Surface Engineering* **16** (2000), 465.
14. Movchan B. A. and Demchishin A. W., *Fis Metall Metallowed* **28** (1969), 653.
15. Thornton J. A., *Ann. Rev. Mater. Sci.* **7** (1977), 239.
16. Zeleny J., *Phys. Rev.* **10** (1917), 1.
17. Rulison A. J. and Flagan R. C., *J Am Ceram Soc* **77** (1994), 3244.
18. Park D. G. and Burlitch J. M., *Chem Mater* **4** (1992), 500.
19. Danek M., Jensen K. F., Murray C. B and Bawendi M. G., *Appl Phys Lett* **65** (1994), 2795.
20. Dobson P. J., Salata O. V. and Hull P. J., *Br. Patent* (1993), 9,323,598.7.
21. Salata O. V., Dobson P. J., Hull P. J. and Hutchision J.L., *Thin Solid Films* **252** (1994), 1.
22. Chen C., Kelder E. M., Van der Put PJJM and Schoonamn J., *J Mater Chem* **6** (1996), 765.
23. Van Zomeran AA, Kelder E. M., Marijnissen J. C. M and Schoonman J., *J Aerosol Sci* **25** (1994), 1229.
24. Unvala B. A., *Br. Patent* (1988), 2,192, 901.
25. Choy K. L. and Su B., *British Patent* (1998), 9900955.7.
26. Vyas J. D. and Choy K. L., *Mater. Sci. & Eng.* **A277** (2000), 206.
27. Choy K. L., Heyes A. L. and Feist J. P., *British Patent* (1998) 9816348.8.
28. Choy K. L., Feist J. P., Heyes A. L. and Su B., *J. Mater. Res.* **14** (1999), 3111.
29. Choy K. L., Mei J., Feist J. P., Heyes A. L., *Surface Engineering* **16** (2000), 469.
30. Choy K. L., Bai W., Chharojrochkul S., Steele B. C. H., *J. Power Source* **71** (1998), 361.
31. Minh N. Q. et al, *in Proc. 3rd Int. Symp. Solid Oxide Fuel Cell* **Vol.93-4** (The Elecrochemical Society, 1993), 801.
32. Yamaguchi R. et al, *in Proc. 3rd Int. Symp. Solid Oxide Fuel Cell* **Vol.93-4** (The Elecrochemical Society, 1993), 704.
33. Van Dieten V. E. J., Walterbos P. H. M. and Schoonman J., *in Proc. 2nd Int. Symp. Solid Oxide Fuel Cell* (The Elecrochemical Society, 1991), 183.
34. Divisek J., DeHaart L. G. J., Holtappels P., Lennartz T. and Mallener W., *J. Power Sources* **49** (1994), 257.
35. DeSouza, Visco S. J. and DeJonghe L. C., *J. Electrochem. Soc.* **144** (1997), L35.

36. Eguchi K., Inoue T., Ueda M., Kaminae J. and Arai H., *in Proc. 2nd. Int. Symp. SOFCs,* ed. Gross F. et al, (Athens, Greece, Commission of the European Communities, Luxembourg, 1991), 697.
37. Tonejc A. M. et al., *Mater. Lett.* **31** (1997), 127.
38. Martinet C. et al., *J. Non-Cyst. Solids* **216** (1997), 77.
39. Boschloo G. K., Goossens A. and Schoonman,J. *J. Electrochem. Soc.* **144** (1997), 1311.
40. Gratzel M., *Common Inorg. Chem.* **12** (1991), 93.
41. D. Mattew, A. Kay and M. Gratzel, *Aust. J. Chem.* **47** (1994), 1869.
42. Choy K. L. and Su B., *J. Mater. Sci. Letter* **18** (1999), 943.
43. Cullity B. D., *Elements of X-ray Diffraction*, 2nd ed. (Addison-Wesley, 1978), 102.
44. Chandrasekhar R. and Choy K.L., *Thin Solid Film* (2001), (in press).
45. Hiroshi et al, *Thin Solid Films* **281** (1996), 202.
46. Chopra K. L. et al, *Thin Solid Films* **102** (1983), 1.
47. Omura K. *et al*, *Renewable Energy* **8** (1996), 405-409
48. Kim S. Y., Kim D. S., Ahn B. T. and Im B. T., *J. Mater. Sci.: Elec. Mater.* **4** (1993), 178.
49. Mooney J. B. and Radding S. B., *Ann. Rev. Mater. Sci.* **12** (1982), 81-101.
50. Adachi A., Kudo A. and Sakata T., *Bull. Chem. Soc. Jpn.* **68** (1995), 3283-88.
51. Sevastian P. J., Campos J. and Nair P. K., *Thin Solid Films* **227** (1993), 190.
52. Su B. and Choy K. L., *Thin Solid Films* **359** (2000), 160.
53. Wu C. and Bube R. H., *J. Appl. Phys.* **45** (1974), 648.
54. Choy K. L. and Su B., *Thin Solid Films* (2001), (in press).
55. Krunks M., madarasz J., Hiltunen L., Mannonen R. and Niinisto L., *Acta Chem. Scandinavica* **51** (1997), 294.
56. Micheletti F. B. and Mark P., *Appl. Phys. Lett.* **10** (1967), 1368.
57. Chavez, Jordan M., McClure J. C., Lush G. and Singh V. P., *J. Mater. Sci.: Mater. Electro.* **8** (1997), 151.
58. Bandic Z.Z., Piquette E. C., McCladin J. O. and McGill T. C., *Appl. Pys. Lett.* **72** (1998), 2862.
59. Shen W. P. and Kwok H. S., *Appl. Phys. Lett.* 65 (1994) 2162.
60. Bonafos C. et al, *Appl. Phys. Lett.* **72** (1998), 3488.
61. Wei M. and Choy K. L., *J. Crystal Growth* (2001), (submitted).
62. Wei M. and Choy K. L., *Advanced Materials* (2001), (in press).
63. Gerenser L. J., Elman J. F., Mason M. G. and Pochan J. M., *Polymer* **26** (1985), 1162.

64. Bachmann M.A. and Koenig J. K., *J. Chem. Phys.* **74** (1981), 5896.
65. Davis G. T., McKinney J. E., Broadhurst M. G. and Roth S. C., *J. Appl. Phys.* **49** (1978), 4998.
66a. Ambrosy A. and Holdik K., *J. Phys. E. Sci. Instrum.* **17** (1984), 856.
66b. Choy K. L. and Bai W., *Thin Solid Films* **372** (2000), 6.
67. Rayliegh J. W., *The Theory of Sound* **2** (1945), 344
68. Gershenzon E. L. and Eknadiosyants O. K., *Akust. Zh.* **10** (1964), 156.
69. Su B. and Choy K. L., *Thin Solid Films* **361** (2000), 102.
70. Barghava R. H., Gallagher D., Hong X. and Nurmiko H., *Phys. Rev. Lett.* **72** (1994), 416.
71. Borelli N. F., Hall D. W., Holland H. J. and Smith D. W., *J. Appl. Phys.* **61** (1987), 5399.
72. Choy K. L., *Mater. Sci. & Eng.* C (2001), (in press).
73. Su B. and Choy K. L., *J. Mater. Chem.* **9** (1999), 1629.
74. Alford N. M., Templeton A. and Penn S. J., *Supercond. Sci. Technol.* **11** (1998), 703.
75. Paranthaman M., Shoup S. S., Beach D. B., Williams R. K. and Specht E. D., *Mater. Res. Bull.* **32** (1997), 1697.
76. Fuenzalida V. M. and Pilleux M. E., *J. Mater. Res.* **10** (1995), 2749.
77. Schmidt J. C., Tigges A. and Schmitz G. J., *Mater. Sci. Eng.* N **53** (1998), 115.
78. Shi J., Berger J. E. and Ling X. S., *Physica C* **301** (1998), 215.
79. Brinker C. J., et al, *J. Non-Cryst. Solids* **63** (1984), 45.
80. Mohideen U., Guuraja T. R., Cross L. E. amd Roy R., *IEEE Transactions on Components, Hybrids and Manufacturing Technology* **11** (1988), 159.
81. Hasegawa K., Nishimori H., Tatsumisago M. and Minami T., *J. Mater. Sci.* **33** (1998), 1095.
82. Su B. and Choy K. L., *J. Mater. Sci. Lett.*, **18** (1999), 1705.
83. McEvoy A. K., McDonagh C. M. and Mccraith B., *Anaylst* **121** (1996), 785.
84. Parrill T. M., *J. Mater. Res.* **9** (1994), 723.

CHAPTER THREE

APPLICATION OF PULSED INJECTION MOCVD TO THE DEPOSITION OF OXIDE SINGLE LAYERS AND SUPERLATTICES

Jean-Pierre Sénateur, Catherine Dubourdieu, V. Galindo, François Weiss
LMGP, ENS de Physique de Grenoble, INPG, UMR CNRS 5628, BP 46, 38402 Saint Martin d'Hères Cedex, FRANCE

Adolfas Abrutis
Vilnius University Dept. of General and Inorganic chemistry - 24 Naugarduko 2600 Vilnius, LITHUANIA

1. Introduction

A large number of oxides are of interest for practical applications: in fact oxides of all transition elements as well as mixed oxide phases have specific functional properties (mechanical and chemical robustness, catalytic, electrical, optical, magnetic…), which may be applied in various areas. Research and investigation into the synthesis and application of these oxides in the form of thin films is continuously increasing. The other problem which needs to be addressed is the formation of integrated multilayer structures combining various layers with different physical properties. These complex materials offer the possibility to tailor magnetic, ferroelectric or superconducting properties by changing the coupling between layers, or to synthesise new materials by association at nanometric scale of compounds which have opposing electronic properties. Oxide superlattice materials, with an artificial control of the crystal structure, are being studied extensively in order to find new and distinct functions from ceramic systems, leading to applications such as powerful piezoelectric actuators, non-volatile memories, IR detectors and Josephson devices. Stress induced effects, related to matching the lattices of two materials, may also strongly influence the desired property.

Until now the main research effort on the synthesis of these layers has been undertaken using physical deposition techniques (PVD) which are generally available for most physicists. The most popular PVD techniques (laser ablation, on-axis or off-axis sputtering, co-evaporation) require, theoretically at least, only solid metal or oxide sources, which are sold by many companies offering chemical compounds. The main difficulty for PVD is the study of the effect of subtle variations of the stoichiometry, which may have a drastic effect on the properties of the layers: a PVD source for a complex oxide deposition is a sintered, homogenous solid plate of material, which is not easy to fabricate, especially when large-area sources are required. It does not make sense, from the point of view of cost, to buy a lot of compounds with only very slight differences in stoichiometry. The problem is quite different for the chemical vapour deposition (CVD) techniques which offer, again theoretically, several potential advantages over PVD deposition such as improved growth rate, ability to coat complex shapes or large areas, and complex investigations in chemical film compositions or in film microstructures because of the easy interchange of source materials, gas flows and temperatures etc. Despite these apparent advantages, the CVD techniques have so far been considerably less frequently used for the deposition of complex oxides than PVD for the following reasons.

- The properties of these layers are of course studied by physicists, who are generally not very confident in studying complex chemical processes.
- For the metal-organic chemical vapour deposition (MOCVD) of complex oxide films[1], the resolution of a number of chemical problems concerning the nature of precursors, the source systems and the chemical interactions between the layer and the substrates is required which, even for a chemist, is by no means simple.

A typical example of the increasing credibility of CVD processing over the last ten years is the progress which has been made on the MOCVD deposition of the $YBa_2Cu_3O_7$ (YBCO) superconducting layer. For practical applications high temperature superconducting (HTS) thin films must meet very high specifications: films must be epitaxial and possess high and uniform critical parameters over all of the deposited surface (critical temperature T_c~90 K, critical current density $j_c>10^6$ A/cm^2 at 77 K for high power applications), and must have a smooth surface and low surface resistance (for microwave applications). The method of synthesis must be able to achieve these high critical parameters, and must also have high reproducibility. This is not a simple task because the quality of films is highly dependent not only on the deposition technique, but also on the structure and properties of the substrate.

Because of the great number of variables in the process, it has taken much longer for the first reports on the successful growth of superconducting YBaCuO layers by CVD to appear in the literature than for other techniques [2]. It is important to note that, in general, chemistry laboratories depend on external physics laboratories for the characterisation of the superconducting properties of their layers, which is a serious drawback. Nevertheless, the quality of high Tc films deposited by CVD has achieved comparability with the best results obtained by physical techniques, but the reproducibility of the superconducting properties was far from equal to the demands of the technical applications. The main problem for the MOCVD deposition of all the complex oxides which are of technical interest is related to the thermal instability and low vapour pressures of some precursors, especially those of Ba, Sr or Rare Earth elements. Effective access to high quality films of these oxides by MOCVD requires volatile and pyrolysable precursors of metal oxides. Compounds having M-O bonds such as alkoxides $[M(OR)_n]_m$ or β-diketonates $[M(P\text{-}dik)_n]_m$ are the classic molecular precursors of metal oxides. So far metal β-diketonates, 2,2,6,6-tetramethyl3-5-heptanedionates (thd) in particular, have been used mainly for MOCVD routes to deposit layers containing these elements. Barium β-diketonates are predominantly oligomers and display medium volatility. Adducts have been obtained in the presence of Lewis bases, such as polyethers, alcohols and amines, and thus with hard donor sites, oxygen or nitrogen, however they often display a poor thermal stability (dissociation in the vapour phase) requiring the addition of the ligand to the carrier gas in order to improve their volatility with respect to the uncomplexed barium derivatives. In addition to a low volatility for barium derivatives, "ageing" phenomena lead to the non-reproducibility of the deposition conditions. These "ageing" phenomena in fact correspond to degradation reactions with loss of volatility and the formation of brown peroxide residues. These observations emphasise the importance of purity (absence of water or oxygen), and of the bonding of the alkaline earth metal derivatives. Molecular chemistry offers the potential to improve the properties of the precursors (volatility, stability) and to build up mixed metal species (associating two metals in a single "source" precursor) which could be another way to facilitate the problems of the transport of barium. A large number of chemistry laboratories are working intensively in this area, but at the moment there is no satisfactory precursor for the conventional MOCVD of Ba, Sr or rare earth. Despite several promising developments in the chemistry of new precursors this problem remains unsolved and continues to impede the technological applications of the MOCVD. This imperfect reproducibility of film properties is a common problem for all deposition techniques.

To summarise, for the MOCVD deposition of complex oxides containing Ba, Sr or Rare Earth element the following problems have to be overcome.

- The precursors used for CVD are chelate compounds which are in the condensed phase at the temperature of evaporation (between 100 and 200°C, they are solid or liquid).
- The concentration of the precursor in the gas phase is generated by thermal evaporation which is strongly dependent on the temperature. The temperature must therefore be controlled within the range of a few tenths of K in order to obtain a constant gas composition.
- The evaporation rate can change with time because of polymerisation or hydrolysis. Therefore the gas composition must be measured during the deposition process to an accuracy of a few %. (but how to do that ?)
- The first experiments on the CVD deposition of complex oxides were carried out with different evaporation sources for each precursor. With thermally unstable precursors, the temperature stability of the source is not sufficient to fix the gas composition which makes the deposition process practically uncontrollable and prevents reproducibility.

In order to make use of the advantages of the CVD process, the problem of reproducibility of the composition and flow rate of the active vapour phase must be solved and consequently thermal evaporation of the solid sources is being studied in several laboratories. The investigation of new injection systems with the high output required for large-scale applications, may result in another way to improve the MOCVD method.

For the deposition of multi-component oxide layers, various physical (e.g. PVD) and chemical methods are being tested. Among the chemical methods, the best quality HTS thin films have been obtained by the MOCVD technique. It is probably the only chemical method which allows *in-situ* deposition of HTS films which exhibit properties identical to those obtained by PVD. MOCVD also has some advantages over PVD: the possibility of achieving higher film growth rate allowing deposition onto large and uneven surfaces. The main problem of classic MOCVD however, is the chemical instability of metal-organic compounds (especially of Ba) during prolonged heating and even during storage. This causes serious problems for the reproducibility of vapour phase composition which is particularly important in the synthesis of multicomponent oxide thin films (e.g., HTS), in which composition must be precisely controlled to obtain films with the necessary physical properties including reproducibility.

2. Deposition of Oxide Films from the Metal-Organic Vapour Phase

Conventional MOCVD for complex materials deposition needs one source for each element. The metal-organic precursors are vaporised by heating in separate evaporators and the temperature of the sources and the flow rate of the vector gas inside the evaporator determine the composition of the vapour phase. The resulting vapour is carried by transport gases towards the substrate, where metal-organic compounds decompose and form the film. The activation energy of the reaction on the substrate can be given by heating, photons, plasma, etc.

For the synthesis of a complex compound using a classic MOCVD source, two independent parameters have to be stabilised for each element in order to control the composition of the gas phase, ie the temperature of the evaporator and the vector gas flow (generally Ar or N_2). The parameters controlling the growth of the layer are threefold, the substrate temperature, the working pressure and, finally, the gas flow in a special line (generally oxygen for oxide deposition). This means that for the conventional MOCVD deposition of a compound containing n elements, 2n+3 independent parameters have to be accurately controlled. In this process, the heating of the organic precursors may cause some decomposition (they are heated during all the successive experiments) which can lead to an uncontrollable drift in the vapour pressure. Even when the precursors are thermally stable at their volatilisation temperature, the stabilisation of the vapour pressure of the different precursors is not easy. Controlling the vaporisation under low pressure for a liquid requires great care but is a real challenge for a solid.

- In a liquid, the bubbling becomes unstable at low pressures.
- In a solid, the evaporation rate depends strongly on the free surface and so fluctuates with the amount of precursor remaining in the source.

Since the first experiments on the MOCVD deposition of high temperature superconductors more than ten years ago, efforts have been made all over the world to alleviate this problem. Alternative single-source approaches are being developed to solve the problem of the vapour phase control and homogeneity for unstable precursors. These techniques, which allow the use of precursors of lower volatilities than for classic multisource MOCVD processes, are all based on the same principle: the greater part of the precursor is maintained in a close vessel at room temperature under inert gas, and a small amount of weighted precursor is steadily introduced into an evaporator held at high temperature, where a flash volatilisation takes place. This has distinctive advantages over the conventional MOCVD sources as follows.

- It avoids the long-term decomposition of the precursors in the tank.
- The time taken to transfer between the evaporator and the reaction zone is very short (sec.), so that the precursor may be overheated without significant decomposition, resulting in very high vapour pressures and thus a very high growth rate (micron/sec) may be achieved (i.e. for ZrO_2), which is impossible with conventional bubblers or sublimators.
- The active vapour flow rate depends entirely on the precursor flow rate inside the evaporator (assuming all precursors introduced are flash-volatilised).
- The evaporator temperature does not require a very high stabilisation.
- The small amount of precursor introduced into the evaporator is flash-volatilised and thus, even with a mixture of precursors, the composition of the vapour phase is the same as that of the mixture so that only one source of known composition is needed for the synthesis of a multi-component oxide. This is one of the major advantages of this technique.

The significant parameters which control the vapour phase are the precursor flow rate inside the evaporator, the vector gas flow rate, and the evaporator temperature. As in the conventional process, three parameters control the growth (temperature, pressure and extra gas flow); this has greatly simplified the MOCVD process since only 6 parameters have to be controlled during the growth for the deposition of a compound containing n elements [(2n+3) if conventional MOCVD sources are used].

The main problem is the accuracy of control of the flow rate of the precursor inside the evaporator. Numerous precursors of elements with strong technological interest (rare earth, Ba, Sr...) are solid both at room temperature and at their volatilisation temperature. Different systems have been proposed where the solid precursor (or mixture of precursors) is injected directly into the evaporator. The main systems which have been used are as follows.

- Direct handling of powders of precursors whereby controlled introduction of the powder of precursors is achieved using a vibrating feeder [3, 4,5].
- Gradual introduction into the evaporator of a compact rod made with a mixture of precursors. This technique has been applied to the deposition of superconducting YBCO. Since the three precursors of Y, Ba and Cu have

very different volatilisation temperatures, the rod is introduced into the evaporator at a very steep thermal gradient (close to 100 °C/mm.)

- Introduction into the evaporator of spherical [6] stainless steel or ceramic, covered by the powder of the precursor. After the evaporation of the volatile precursor, the spheres are recycled.

These systems give very good results on a laboratory scale. However, it is not a simple task to produce large quantity of homogeneous rods with very fine grains. This is because of the difficulties in direct handling and accurate feeding of such powders which are either sticky (if wet), or electrostatic (if dried).

These systems give very good results on a laboratory scale, but the realisation at a large scale of homogeneous rod of sticky (if wetted), or electrostatic (if dried), or the direct handling of such powders with a high accuracy of feed rate, is not a simple task. This high degree of accuracy of the feed rate is a key parameter in the synthesis of materials composed of several chemical elements with single source systems. The different precursors of a material of high technological interest generally have different volatility and thermal stability levels, and the deposition is very far from congruent (see Fig. 4). The relative reaction yield for the deposition of the different elements changes with the deposition parameters (deposition temperature, pressure, flow rates), which means that a constant flux of the precursor is a condition *sine qua non* for obtaining homogeneous, high quality samples.

The control of the flow rate is considerably easier for a liquid than for a solid and a large number of solutions have been suggested where the precursor, or a weighted mixture of precursors is dissolved in an organic solvent. This solvent is selected because it does not change the chemistry of the precursor, and is easily vaporised (diglyme, benzyl alcohol, monoglyme, tetrahydrofuran, hexane). Different techniques have been proposed for the introduction of the solution into the evaporator.

- Aerosol carried by a vector gas [1, 7, 8, 9, 10]. The aerosol is generally produced by ultrasonic excitation of the liquid, the droplets are very fine (few µm.), with a narrow distribution of size. The size of the droplets is fixed by the frequency (several kHz), viscosity and density of the solution. Two different techniques have been used to produce the aerosol: a piezo electric ceramic in the bottom of a reservoir, or injection of the liquid by means of an ultrasonic nozzle. In both techniques, a rather high energy is transmitted to the solution and to the transducer, which have to be carefully thermostated. The main problem is the stabilisation of the flux of aerosol.

- Micro pump, a syringe driven by a motor or liquid mass flow controllers [11, 12].

In all these systems, a slow flow of liquid has to be transferred from the zone where the flow rate is controlled (maintained at room temperature), to the evaporator at high temperature (100-300° C), under low pressure. Generally this transfer is made through a capillary tube which limits the direct contact between the liquid and the low pressure inside the evaporator (a direct connection will lead to the volatilisation of the solvent). This capillary tube is the great weakness of all these systems since it could become blocked by the precipitation of solid precursor if some part of the solvent is volatilised in the zone where there is a thermal gradient. For industrial equipment, more sophisticated systems are used to try to alleviate this problem such as injection by a three-way valve which allows the introduction of a mixture of vector gas and solution inside the capillary (ASM); capillary injection through a hot porous stainless steel element (ATMI); or the very complex design of a multi-disk evaporator (MKS).

At LMGP we have developed, in collaboration with Vilnius University, a very simple liquid delivery system for CVD sources based on the principle used for the control of fuel injection in the latest thermal motors [13]. In our source system the solution is maintained at room temperature in a container connected to a high speed micro electro-valve. Inert gas pressure (typically 1 Bar of argon) is applied to the liquid inside the container, which pushes the liquid towards the valve and when the valve opens the liquid is ejected at high speed. The flow rate is controlled simply by successive fast openings of the micro electro-valve, as in the fuel injection system. The two parameters which control the flow rate are the opening time of the valve (typically 1 to 10 milliseconds), and the frequency of the injection (typically 1 Hz in most of our experiments). These two parameters are defined by electrical pulses driven by a computer, which gives a high degree of flexibility to the process. The minimum width of the electrical pulses is 1 millisecond, which roughly corresponds to 0.2 to 4 mg of injected (depending on the kind of injector used).

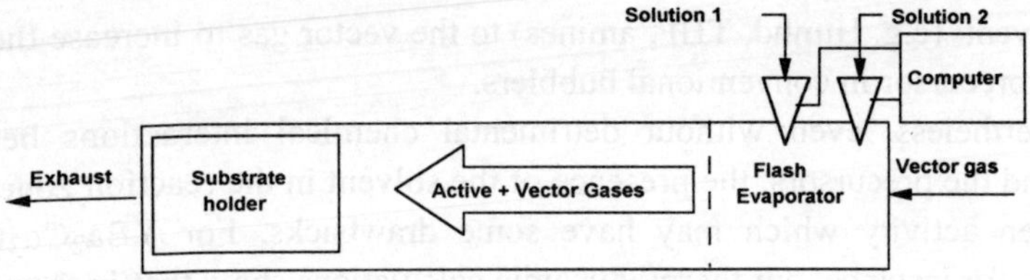

Fig 1 Schematic representation of an injection CVD reactor with two injectors.

The reproducibility of the system is better than 3%, and a very large range of flow rates are available by adjusting both the frequency (1 to 100 Hz) and the pulse width (2 to 8 milliseconds), and to control the flow rate between 0.05 and 100 g./min. Another peculiarity of the process is that it works "digitally": the thickness may be simply controlled by the *number* of drops injected, which is very interesting for multilayers (for instance, for the deposition of superlattices we adjust the concentration of the solutions and the opening time to obtain a mean thickness deposited at each injection less than 0.1 nm). This system (Figs. 1 and 2) has been successfully applied to the deposition at laboratory scale of a wide variety of simple or complex compounds [14], and has also been used for the deposition of multilayers or superlattices [15]:

- Al_2O_3, Ta_2O_5 [16], SiO_2 [17], MgO, TiO_2 [18], Y_2O_3, $ZrO_2(Y)$ [19], CeO_2 [20] Ta_2O_5-SiO_2 [5], $YBa_2Cu_3O_{7-x}$ [21, 22, 23, 24, 25]
- $YBa_2Cu_3O_{7-x}$ / $PrBa_2Cu_3O_{7-x}$ [26]
- $BaTiO_3$, $SrTiO_3$, $Ba_{(1-x)}Sr_x TiO_3$ and $BaTiO_3/SrTiO_3$ superlattices [27]
- $BaFe_{12}O_{19}$ [28, 29]
- $La_{(1-x)}MnO_3$ [30, 31, 32, 33], $La_{(1-x)}Sr_xMnO_3$ [34], $Nd_{(1-x)}Sr_xMnO_3$ [35]
- Ti(C,N)
- Cu
- $YBa_2Cu_3O_{7-x}$ / $ZrO_2(Y)$//Metal [36, 37]
- $YBa_2Cu_3O_{7-x}$ / $SrTiO_3$//$LaAlO_3$

Pure metal-organic precursors contain a large amount of carbon and the presence of the solvent in the reaction zone increases by one order of magnitude the ratio of carbon: metal in the gas phase. We have not, however, found any significant difference in the carbon concentration of the layers deposited using either conventional or liquid-injection CVD sources (NRA, SIMS, SNMS, XPS or electronic properties). In some cases, the solvent has a favourable effect on the stability of the vapour pressure of the precursors: some authors have systematically added solvent (e.g. Htmhd, THF, amines) to the vector gas to increase the volatility of the Ba precursor in conventional bubblers.

Nevertheless, even without detrimental chemical interactions between the solvent and the precursors, the presence of the solvent in the reaction zone decreases the oxygen activity which may have some drawbacks. For $YBa_2Cu_3O_{7-x}$ layer deposition for instance, our thermodynamic calculations show that in the presence of carbon, where the oxygen activity is too low, $YBa_2Cu_3O_{7-x}$ is not stable. High growth rate of the layers needs high solution feeding rates, which must be compensated for by increasing the oxygen partial pressure. This increase of oxygen partial pressure in a gas phase containing a large amount of organic solvent at 800 °C, leads to an explosive situation. In fact higher growth rates may be obtained without solvent in the reaction zone.

Three solutions have been proposed involving the use of liquid sources to facilitate the control of the flow rate of the precursor, but with elimination of the solvent from the reaction zone. The first technique [38, 39], is based on the *ex-situ* wetting of a porous fibreglass tape. The dried tape is then carried mechanically inside the evaporator, where the precursor is flash-volatilised. Other systems use *in situ* wetting and separation of the solvent[40]. A schematic view of our system is presented in Fig 2.

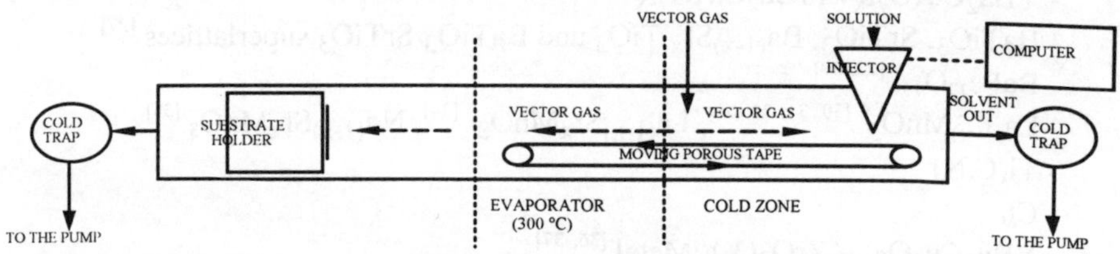

Fig 2. Schematic representation of an injection CVD reactor with *in-situ* elimination of the solvent. Only one injector is represented, but this system also works with several injectors, allowing the synthesis of multilayers. The drops are injected in the "cold zone" under low pressure, the solvent volatilises and the gas flowing towards the right carries away its vapours. The moving tapes carry the dried precursor into the evaporator, and the vapours are carried by the gas flowing toward the substrate.

3. Experimental Set-Up and Characterisation

Perovskite layers were deposited on single crystal MgO (001), YSZ (001), $NdGaO_3$ (001), $LaAlO_3$ (001), sapphire (1102), $SrTiO_3$ (001), Si (001), and on stainless steel substrates.

The critical temperature (T_c) of the $YBa_2Cu_3O_{7-x}$ layers was determined by resistance vs. temperature measurements using the standard four-probes technique, and from contactless AC susceptibility measurements.

Critical current density (J_c) was determined by voltamperometric measurements in a narrow bridge, using short (10-15 ns) current pulses, and AC susceptibility third harmonic measurements.

For the study of film surface morphology, optical microscopy, scanning electron microscopy (SEM) and atomic force microscopy (AFM) were used.

Film microstructure was investigated by X-ray Diffraction (XRD) in Bragg-Brentano and Schulz geometry, transmission electron microscopy (TEM) and high resolution transmission electron microscopy (HRTEM)

The film composition was determined by energy-dispersive X-ray analysis (EDX) operating with SEM; the calibration of the EDX has been made by RBS on some selected samples. X-ray photoelectron spectroscopy (XPS) and secondary ion mass have been also used for the determination of the homogeneity of the layers and of the inter diffusion at the substrate's interface.

Film thickness was measured by optical interference microscopy, RBS, profilometry, ellipsometry and optical (laser) reflectometry (for *in-situ* measurements).

Magnetisation vs. temperature measurements was performed with a vibrating sample magnetometer. Magnetoresistance vs. temperature was measured in the fields up to 5 T by the standard four-probe technique.

4. Results

4.1. Oxide films grown by pulsed injection CVD

4.1.1. Synthesis of $YBa_2Cu_3O_{7-x}$ films by pulsed injection CVD

The optimised deposition conditions under which high quality YBCO layers were reproducibly grown, are given below. Under such conditions the YBCO growth rate is ~2 μm/h, i.e., about 5-10 times higher than that achieved in most classical CVD reactors.

The optimum deposition conditions for YBCO by injection-MOCVD are:

Substrate temperature	825 °C
Evaporator temperature	290 – 310 °C
Total pressure (Ar + O_2)	6.6 hPA
Oxygen partial pressure	2.6 hPA
Solvent	Monoglyme
Solution concentration	0.02 Mol/l (Y(thd)$_3$)
Pulse time	2 ms
Injection frequency	1 Hz
Mass at each injection	5 mg

A systematic study was performed in order to determine the influence of the main deposition conditions (substrate temperature and vapour composition), and of the nature of the substrate, on YBCO film properties. This systematic study allowed us to gain a better control of the quality of YBCO films. The results obtained are presented below.

4.1.1.1. YBCO films on YSZ (001) substrates

YSZ and YBCO have different crystal structures and their lattice mismatch is rather large - about 6% in (001) and so it is difficult to deposit epitaxial YBCO layers with good critical parameters. The properties of films on YSZ substrates depend strongly on the deposition technique used, and on the deposition conditions. There are several reasons for interest in the study of YBCO deposition on YSZ: (i) this substrate is cheaper than perovskite substrates, (ii) YSZ is widely studied as a potential buffer layer for YBCO deposition on sapphire, silicon, or metallic substrates, and these studies may be partly based on the data obtained from YBCO deposition on YSZ single crystal substrates.

YBCO thin films (0.3-0.4 μm) were grown at various substrate temperatures and a wide range of vapour compositions. Layers with the best critical parameters were obtained at the solution composition Y:Ba:Cu = 1:2:2.5 (at %). YBCO films deposited at 850-875°C were strongly c-axially orientated (c⊥). Films deposited at 825°Cexhibited weak (*k*00) reflections in XRD θ/2θ spectra, which showed that they contain a small amount of YBCO crystallites with an a-axis orientation (a⊥). The amount of a-axially orientated crystallites increases with the decrease in the deposition temperature. All films deposited within the range 750 - 875°C have CuO precipitates on their surface. Their size increases and their density decreases when the substrate temperature rises.

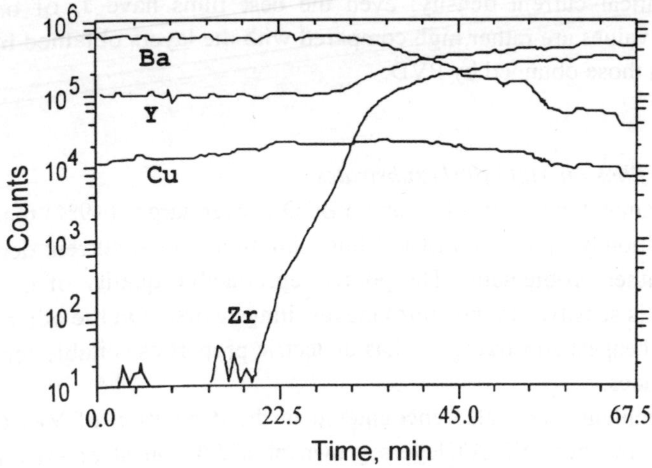

Fig 3. SIMS spectra of YBCO deposited at 850 °C.

The critical temperature of YBCO films clearly depends on the substrate temperature. At 775°C, the deposited films exhibit resistivity vs. temperature dependence similar to a semiconductor. The best films were obtained at 825°C [$T_c(R=0) = 89 - 90$ K, $\Delta T_c = 1-1.5$ K] and the properties are slightly worse at 850°C, films deposited at 875°C have a considerably lower T_c (~82 K), and those grown at 900°C do not exhibit a transition to the superconducting state until above 77 K. The degradation of the critical temperature with the increase in growth temperature might be caused by the interaction between YSZ substrate and YBCO film. The SIMS (secondary ion mass spectrometry) profiles reveal a rather deep Zr diffusion into the YBCO layer even at 825-850°C (Fig. 3), and in addition XRD θ/2θ spectra for these films show weak $BaZrO_3$ reflections.

The composition of the vapour phase strongly influences the in-plane orientation of the films. YBCO crystallites in the films have several different orientations but the 45° orientation ([100]YBCO//[110]YSZ]) is always more or less predominant. In the best layers, the 45° orientation dominates and only a very small amount of 0°-orientated crystallites could be seen in XRD φ-scans; however the full widths at half maximum (FWHM) of φ-scan peaks and rocking curves are rather large (4-5° and ~0.9° respectively), which shows that the films are not completely epitaxial. This may be caused not only by the rather large lattice mismatch between YSZ and YBCO, but also by the formation of a non-orientated intermediate $BaZrO_3$ layer at high growth temperature. The lower epitaxial growth of YBCO leads to a rather low critical current density: even the best films have J_c of only 4-5x10^5 A/cm^2. These values are rather high compared with the layers obtained by MOCVD, but lower than those obtained by PVD.

4.1.1.2 YBCO films on MgO (001) substrates

The lattice mismatch between MgO and YBCO is even larger (~9%) than with YSZ and the deposition of epitaxial YBCO films with high critical current density on this substrate is rather problematic. The poorly reproducible quality of the substrate's surface (MgO is sensitive to moisture) makes things worse. On the other hand, MgO is one of the cheapest substrates and has dielectric properties suitable for use in high frequency devices.

These problems were also encountered in the deposition of YBCO layers by PICVD. Even the best YBCO layers grown at 825°C on standard (commercial) MgO substrates do not exhibit high critical parameters: T_c~86-87 K, ΔT_c~2-3 K, J_c~10^5 A/cm^2 (77 K). This critical current density is lower than the best values reported for YBCO obtained using conventional sources. We have no explanation of this poor performance, but it is possible that the presence of the solvent increases the concentration of CO, CO_2, H_2O in the gas phase, which reacts with the MgO surface. On the other hand, the poor reproducibility of the conventional MOCVD for the synthesis of YBCO layers leads, in 90% of the samples, to much lower performances. PICVD layers are at least reproducible. The disappointing properties are due to the microstructure of the films which are strongly c-axially orientated, however in the substrate plane YBCO crystallites are orientated in several different directions with 0° predominating ([100]YBCO//[100]MgO). The presence of high-angle intergrain boundaries causes low critical current densities. The attempt to obtain pure in-plane orientation in YBCO films on standard MgO substrates by varying deposition conditions (temperature, vapour composition), was unsuccessful.

It was expected to improve YBCO thin film properties using vicinally-polished MgO substrates. The surface was polished with an inclination of 1.4-1.9° from the (001) plane along [100] direction. Using these substrates, the influence of the vapour phase composition on the composition, surface morphology, in-plane orientation and critical parameters of the YBCO layers was investigated. Layers of thicknesses close to 0.2 μm were deposited at 825°C. The solution composition was varied over a wide range (Y from 17 to 25, Ba from 35 to 45 and Cu from 30 to 40 at %).

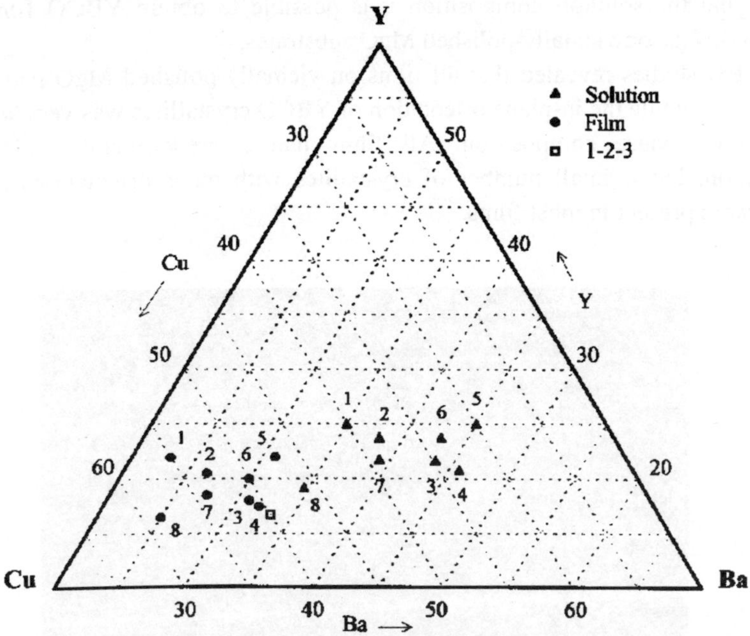

Fig 4. Composition (metal atom %) of both the solution and the layers : note the shift related to the different reaction yields of the precursors

All films were deposited from the vapour phase enriched in Y and Ba and poor in Cu comparison with the stoichiometric 123 composition (Fig. 4). At the same time the films had an excess of Y and a deficit of Ba compared with 123 composition, while the proportion of Cu varied at ±50%. In comparing vapour phase and film compositions, it was found that at Ba and Y atomic percentages the films were lower by a factor of about 1.2 and 1.5 respectively, compared with the vapour phase, while the percentage of Cu in the film was higher by an average factor of about 1.5. This effect was related to the different decomposition rates of the precursors.

The SEM images of the surface of the films revealed a smooth film matrix, but various defects were present in most of them. The density and the type of defect depended on vapour (and film) composition. A local EDX analysis showed that the inclusions of additional phases had compositions close to those of Y_2BaCuO_5,

$Y_2Cu_2O_5$, and CuO phases. Films with composition close to the 123 stoichiometry had a smooth surface without inclusions (Fig. 5). These investigations showed that by varying the solution composition it is possible to obtain YBCO films with a smooth surface on vicinally-polished MgO substrates.

XRD studies revealed that all films on vicinally-polished MgO had a pure c-axis texture while the in-plane orientation of YBCO crystallites was very sensitive to the vapour phase composition. All films had a predominantly 45° in-plane orientation, but a small number of crystallites with other orientations (0°, ~15°, ~40°) were present in most films.

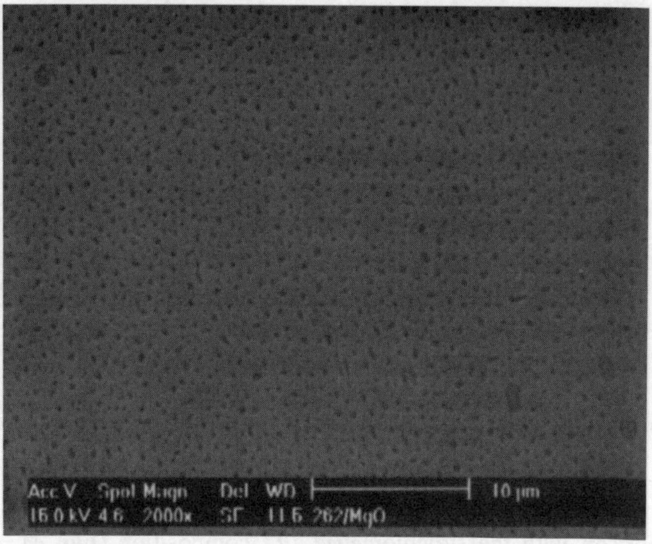

Fig 5. SEM image of YBCO film on vicinal MgO. The composition of this film is close to 1-2-3.

Only the films with average composition close to the stoichiometric 123 had pure 45° orientation, which is very different from that generally observed on MgO where the 45° orientation generally did not exist, or appeared as a minor phase[41]. The epitaxial relationship between the substrate and the film can be written as (001)[110]YBCO// (001)[100]MgO (this is not absolutely accurate because the vicinal angle is not taken into consideration).

The critical temperatures of the films on MgO were not noticeably dependent on the vapour composition (highest T_c = 89 K, ΔT_c = 1.5 K). The critical current densities varied widely depending on the in-plane orientation of the films. The highest J_c values (>10^6 A/cm^2 at 77 K) were found in the films with a pure (45°) orientation.

Such high values (>10^6 A/cm^2 at 77 K) are comparable to the best obtained on MgO substrates by PVD.

4.1.1.3 YBCO films on $LaAlO_3$ (001) substrates

$LaAlO_3$ has a rhombohedral structure, a slight distortion of a perovskite structure. The match of its lattice parameters and thermal expansion coefficients to those of YBCO is good and therefore it is quite easy to deposit epitaxial thin films on this substrate. Usually, the reported J_c values (at 77 K) are 2-3 10^6 A/cm^2, but there were some reports about higher J_c values of up to (5-6) x 10^6 A/cm^2 obtained by PVD and plasma-enhanced CVD.

YBCO films (~0.2 μm) were deposited by PICVD on $LaAlO_3$ at 825°C from vapour phases with various compositions. The relationship between vapour phase and film composition revealed a trend consistent with that of other substrates: the films had relatively less yttrium and barium, but more copper than in the vapour phase. All deposited films were enriched in yttrium and poor in barium, compared with the 123 composition, while the proportion of copper varied at ±50% of the stoichiometric amount (as with the films on MgO).

A pure c-axis texture was found in most deposited YBCO films. Films had perfect 0° in-plane orientation on a pseudo-cubic $LaAlO_3$ R-plane. The FWHM of (005) rocking curves for these films was very small - 0.25-0.45°. These values, together with clearly observed YBCO twinning in two directions ([110], [1$\underline{1}$0]) demonstrated the very high epitaxial quality of the films, however with the increase in the amount of barium in the vapour phase, a small number (<10%) of a-axis orientated YBCO crystallites appeared in the films, as is shown in the SEM image presented in Fig. 6.

The critical temperatures of the films were almost independent of vapour composition. All deposited films had a T_c of about 90-91 K (R=0) and a ΔT_c of 0.31 K. The films exhibited good metallic properties with a room-temperature resistivity of 220-280 $\mu\Omega$cm and a resistivity ratio R(300 K) /R(100 K) close to 3.0. The critical current density was only slightly dependent on the vapour composition in the range studied (in the same region as shown in Fig 4). Most films had very high J_c values - 6-7 x 10^6 A/cm^2 (77 K). Films with a small number of a$\perp$ crystallites had slightly lower J_c values of about 5 10^6 A/cm^2, which is comparable to the state-of-the-art [42,43,44,45,46]. Only one layer, with the lowest average amount of copper, clearly had a lower J_c value ~1 x 10^6 A/cm^2.

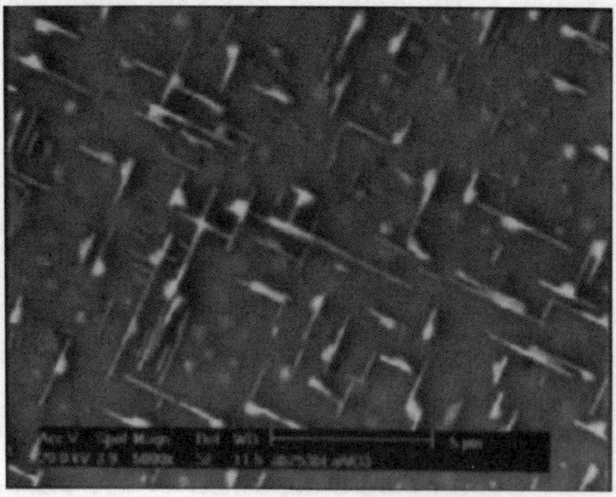

Fig 6. SEM image of YBCO film on $LaAlO_3$. The needles are a-axis crystallites.

4.1.1.4 YBCO films on $NdGaO_3$ (001) substrates

$NdGaO_3$ has an orthorhombic perovskite type of structure, in which the lattice parameters match very well those of YBCO. J_c values of 1-2 x 10^6 A/cm^2 at 77 K are usually reported for YBCO films on this substrate but some have reported higher values of (2-3) x 10^6 A/cm^2. The lower J_c values obtained on $NdGaO_3$ in comparison with films deposited on $LaAlO_3$, are due to the presence of more a-axially orientated crystallites in most films on $NdGaO_3$.

Mixed c-axis and a-axis orientated crystallites were also found in all YBCO films deposited on $NdGaO_3$ by PICVD. The ratio of the two orientations depended on the vapour phase composition (Table 1). The percentage of a⊥ crystallites increased with an increasing amount of Cu and especially Ba in the vapour phase (the same tendency was also observed in films on $LaAlO_3$ and YSZ substrates). The largest proportion of a⊥ crystallites was obtained in the films with a composition close to the stoichiometric 123, and the proportion of a⊥ crystallites was greater in thick films deposited under similar conditions. Film surface morphology was influenced by the c⊥ and a⊥-texture ratio in the films. In films with predominantly c⊥-texture, a small quantity of a-orientated crystallites was visible on the smooth surface. With increasing proportion of a⊥-oriented crystallites in the film, the density of these crystallites on the surface increased until they covered the whole surface (Fig. 7) and formed a typical surface picture of a⊥ texture. In all films the in-plane axes of a⊥ and c⊥ crystallites were orientated at 45° with respect to the substrate axes, which in both cases corresponds to the best matching of lattice parameters. Both (110) and (1$\underline{1}$0) twin planes were clearly observed in the X ray phi scans for the *c*⊥-oriented domains of all deposited films.

Table 1. Properties of YBCO films, in relation to the vapour phase composition.

Y : Ba : Cu in solution	a perpendicular crystallites %	Tc (DTc) K	Jc (77 K) (10^6 A/cm^2)
1 : 1.4 : 1.6	11	NO SUPRA	
1 : 1.6 : 1.4	6	89.8 (0.3)	5.2
1 : 1.6 : 1.6	18	89.8 (0.4)	3.0
1 : 1.6 : 1.8	25	89.5 (0.6)	2.0
1 : 1.8 : 1.2	3	NO SUPRA	
1 : 1.8 : 1.4	5	90.2 (0.4)	4.2
1 : 1.8 : 1.8	20	90.1 (0.6)	2.1
1 : 1.8 : 2.0	33	89.2 ((0.7)	1.8
1 : 1.8 : 2.2	38	89.7 (0.6)	1.5
1 : 1.8 : 2.4	41	88.9 (0.8)	1.2
1 : 2.0 : 1.6	56	89.4 (0.5)	0.9
1 : 2.0 : 1.8	43	89.7 (0.5)	1.0
1 : 2.2 : 1.6	54	89.6 (0.6)	0.8
1 : 2.2 : 1.8	71	89.8 (0.8)	0.5
1 : 1.9 : 2.2	93	90.2 (0.5)	0.2

YBCO films on $NdGaO_3$ exhibited a high T_c ~ 90 K and a sharp superconducting transition (Table 2). T_c of the films was almost independent of the vapour phase composition in the large range studied, and therefore on the ratio of a⊥ and c⊥ orientations in the films.

The ratio of the two orientations strongly influenced the J_c values of the films and J_c decreased with a higher proportion of a⊥-oriented crystallites (Fig. 8).

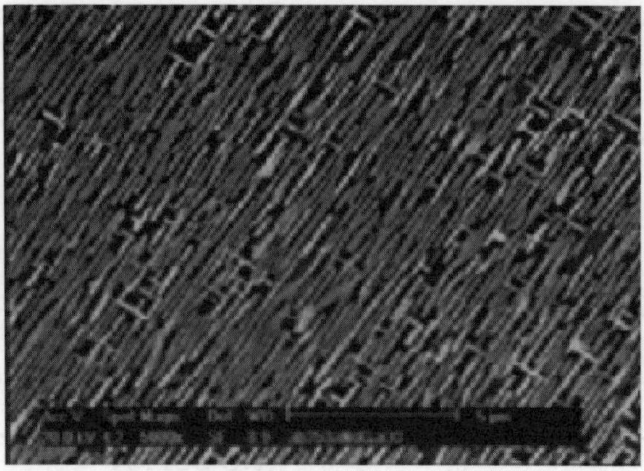

Fig 7. SEM image for a YBCO film on $NdGaO_3$ showing 56% of a-axis crystals.

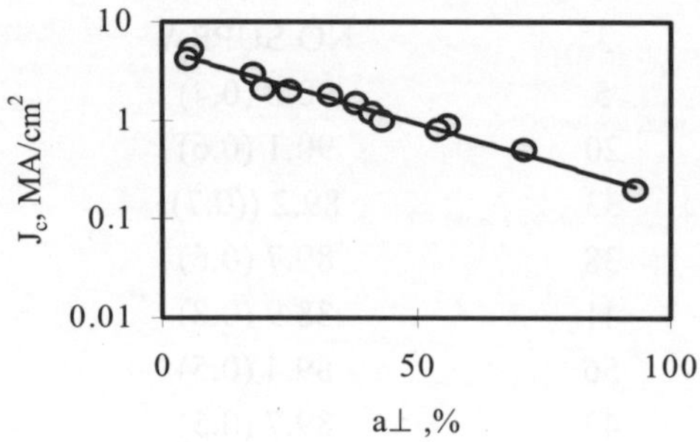

Fig. 8. J_c (77 K) dependence on the quantity of a-axially orientated crystallites in YBCO films on $NdGaO_3$.

4.1.1.5 YBCO films on sapphire (1̲102) substrates with a CeO_2 buffer layer

Sapphire is a very desirable substrate for the deposition of HTSC layers because of its technological features. However, the significant lattice mismatch (~6% for the R plane), and the chemical interaction between Al_2O_3 and YBCO at high temperatures, make the YBCO deposition very problematic and the considerable mismatch between thermal expansion coefficients of sapphire and YBCO causes cracks in films thicker than 400 nm. One of the methods employed to solve these problems was to deposit a suitably thin buffer layer prior to YBCO deposition.

CeO_2 was chosen as a buffer layer for YBCO synthesis by PICVD. The parameter of the cubic CeO_2 lattice is 5.41 Å, and its match with the parameters of sapphire is not very good (~5.7% with R plane), nevertheless, this oxide is a good anti-diffusion barrier for YBCO on sapphire. On the other hand, CeO_2 is a good substrate for the epitaxial growth of YBCO (the lattice mismatch is only 0.9%). The main problem encountered in the deposition of CeO_2 layers by various techniques is the difficulty in obtaining a pure (001) orientation of the CeO_2 layer without inclusions of (111)-orientated crystallites. The presence of these inclusions lowers the quality of the deposited YBCO film.

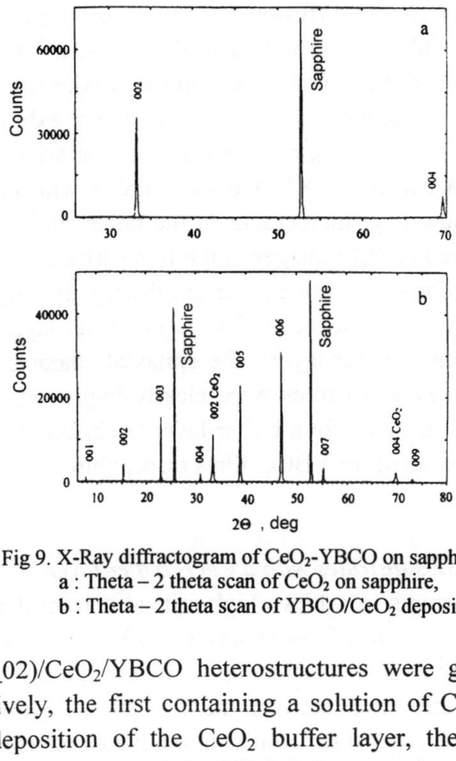

Fig 9. X-Ray diffractogram of CeO_2-YBCO on sapphire
a : Theta – 2 theta scan of CeO_2 on sapphire,
b : Theta – 2 theta scan of YBCO/CeO_2 deposited in-situ on sapphire.

Sapphire(1$\underline{1}$02)/CeO_2/YBCO heterostructures were grown *in-situ* using two working successively, the first containing a solution of Ce(thd)$_4$ in heptane (0.02 mol/l) for the deposition of the CeO_2 buffer layer, the second containing the convenient mixture of Y, Ba and Cu (all thd) in monoglyme (0.02mol/l Y (thd)$_2$ for the deposition of the superconducting layer. Immediately after the deposition at 850 °C of a thin (~0.2 μm) CeO_2 layer, the YBCO layer (0.6-1 μm) was grown at 825°C. A solution of Ce(thd)$_4$ in heptane was used for the CeO_2 deposition.

CeO_2 thin films grown at 800°C had some (111) orientated inclusions in the (001) orientated matrix. Traces of these inclusions were also found in films deposited at 825°C. At a deposition temperature of 850°C, the CeO_2 films and the CeO_2 in the heterostructures had a pure (001) orientation (Fig. 9). YBCO layers in such heterostructures had a c-axis orientation. Low values of FWHM for CeO_2 (002), and YBCO (005) rocking curves (~0.8° and ~0.6° respectively), demonstrated a low mosaic spread in both films. XRD study of the deposited heterostructures in Schulz geometry revealed a good in-plane orientation for both CeO_2 and YBCO layers.

SIMS concentration profiles for YBCO/CeO_2/Al_2O_3 structures revealed that the CeO_2 buffer layer stopped the Al diffusion from the substrate into the YBCO film. Consequently, the epitaxial quality and critical parameters of YBCO films in heterostructures were considerably improved compared with those of YBCO films deposited directly on to sapphire. The heterostructures exhibited high critical temperatures (T_c ~90 K, ΔT_c ~0.7 K), but the values of the critical current densities were rather low (~10^5 A/cm^2 at 77 K), and comparable with previously reported values. This could be caused by microcracks in the rather thick (0.6-1 μm) YBCO layers which were observed in SEM images of the film surface.

We attempted to increase the critical current density by varying the thickness of the CeO_2 buffer layer and the thickness of the CeO_2 buffer layer was decreased to 15-100 nm. In these thinner buffer layers, the epitaxial characteristics and critical current densities of the heterostructures were clearly improved. The best J_c value obtained was ~10^6 A/cm^2 at 77 K for a buffer layer thickness of about 30 nm. This value is among the best reported for YBCO films on sapphire.

4.1.1.6 YBCO films on YSZ substrates with a CeO_2 buffer layer

As mentioned above, attempts to achieve high critical current densities for YBCO films on YSZ substrates by PICVD were unsuccessful. The main reason was Zr diffusion from the substrate into the film and we attempted to solve this problem by depositing a thin CeO_2 buffer layer between the YSZ substrate and the YBCO film. Both oxides have the same type of structure, but the lattice matching between CeO_2 and YSZ is not very good (4.6%). Nevertheless, some reports on the possibility of growing epitaxial CeO_2 layers on YSZ can be found in the literature.

YSZ/CeO_2/YBCO heterostructures in a PICVD reactor were grown *in-situ*. First, a CeO_2 layer (~0.1 μm) was deposited at 850°C, using a solution of Ce(thd)$_4$ in heptane. The YBCO layer (~0.3-0.4 μm) was deposited at 825°C.

As with sapphire substrates, CeO_2 layers on YSZ had no (111) orientated inclusions. Diffractograms showed no reflections of the $BaZrO_3$ phase which was found in YBCO films grown directly on to YSZ substrate. Also, contrary to some literature sources, there was no evidence for the chemical interaction between CeO_2 and YBCO. SIMS concentration profiles showed that even a thin CeO_2 buffer layer prevents Zr diffusion into the YBCO layer (Fig 10).

In CeO_2/YBCO structures both CeO_2 and YBCO layers were perfectly orientated in the substrate plane with the following epitaxial relationship (001)[001]YSZ//(001)[001]CeO_2//(001)[110]YBCO. Another feature of YSZ/CeO_2/YBCO heterostructures was the evident YBCO twinning. Two separated peaks can be seen in XRD φ-scans for the 407 reflection of YBCO on CeO_2/ZrO_2, while only a broad peak is seen for the same reflection for YBCO on ZrO_2, confirming that the surface chemical interaction angles differ by 90°. The twinning of the structure is noted along two equivalent directions [110] and [1$\underline{1}$0].

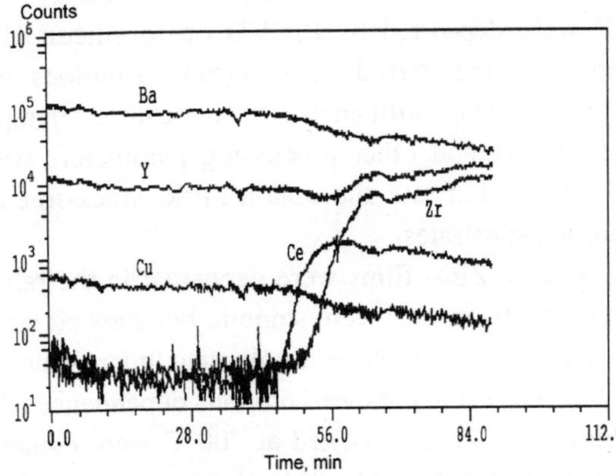

Fig 10. SIMS concentration profile of a ZrO_2/CeO_2/YBCO heterostructure. The thickness of the CeO_2 layer is 30 nm.

Evident twinning was also observed in epitaxial YBCO layers on $NdGaO_3$ and $LaAlO_3$ substrates. Twinning was less clear in the films on vicinally polished MgO and not visible at all in the layers deposited directly on to a YSZ substrate. The critical parameters of the heteroepitaxial YSZ/CeO_2/YBCO structures obtained were very high, T_c ~90 K, ΔT_c ~0.3-0.4 K, J_c ~4 x 10^6 A/cm^2 at 77 K; compared with the layers on YSZ, J_c increased by an order of magnitude and this might be related to the considerable improvement of the YBCO film microstructure. A variation in the thicknesses of the CeO_2 buffer layers in the range of 15-100 nm had no noticeable influence on the properties of the YBCO layers which were consistently high.

These results demonstrate that simple *in-situ* deposition of a CeO_2 buffer layer allows one to obtain YBCO films on YSZ of a quality comparable to that on the best perovskite substrates.

4.1.2 ZrO_2 and YSZ films

Pure ZrO_2 or yttrium-stabilised ZrO_2 films are extensively investigated for numerous applications in optics and electronics, hard coatings, thermal barriers and ionic conductors. Because of its high melting point, hardness and chemical inertness, zirconium oxide is interesting as a material for protective coatings. YSZ films can be used as a solid electrolyte in oxygen sensors and fuel cells. Moreover, YSZ is one of the most promising buffer layers for HTSC deposition on silicon and other technologically important substrates such as sapphire or metals.

ZrO_2 and YSZ were deposited by PICVD on to silicon (100) and polished stainless steel substrates using $Zr(thd)_4$ and $Y(thd)_3$ solutions in tetrahydrofuran (THD) or monoglyme. The influence of deposition temperature, yttrium concentration, gas flow rate and other processing parameters was investigated in order to increase the deposition rate and obtain thick, crack-free coatings on these technologically important substrates.

On Si/SiO2 substrates, ZrO_2 films were deposited in the range of 700-900°C. Films obtained at lower temperatures were smooth, but showed large cracks in SEM images. Films deposited at >750°C have a columnar microstructure, and show no evidence of cracks. The XRD study showed obvious dependence of film structure on the deposition temperature. Films deposited at 700°C were composed of very fine grains of the tetragonal/cubic phase. At substrate temperatures of 800°C and above, the monoclinic phase became clearly evident in the films. Both the low and high temperature phases had a preferred (001) texture. The stabilisation of the cubic high temperature (>870 °C) phase of pure ZrO_2 could be observed at deposition temperatures as low as 700 °C is explained by an increase of the surface energy related to very small grain size. YSZ films deposited at 900°C on silicon and stainless steel substrates had good adhesion to the substrate. The influence of yttrium addition on the stability of monoclinic structure was investigated. At this deposition temperature, the structure transformation monoclinic-(tetragonal/cubic) started at a very low yttrium concentration in solution (2.5 at % in the Zr+Y mixture) and with the increase in percentage of yttrium content, the cubic phase became more dominant and at 10 at% the monoclinic phase disappeared. All YSZ layers on Si had microcracks, which became more and more evident with increasing Y concentration and film thickness. In YSZ films on stainless steel substrates, microcracks appeared only in the case of Y quantity >10 at% and thickness >1.5 μm, but even thick films remained very adherent in contrast with films deposited on silicon.

An important factor in the deposition of YSZ thin films for many technological applications (thermal barriers, hard coatings, electrolytes) is the growth rate. The study of the influence of the processing parameters on film growth rate showed the conditions under which YSZ films could be grown at the rate of 27 μm/h. Such a rate is very high, compared with those achieved using other techniques, and is impossible to obtain using classical evaporators or sublimators for CVD.

One of the main results of these investigations was that they established the potential to deposit thick crack-free and adherent YSZ layers on stainless steel at high deposition rates using PICVD.

4.1.3. Ta_2O_5 films

Ta_2O_5 films are applied in microelectronics, optics and as protective coatings. These applications are based on the high dielectric constant (ε_r=25), high refractory index (2.2-2.4) and chemical inertness of tantalum oxide. Ta_2O_5 films have already been deposited by MOCVD using different precursors, but mainly with $Ta(OEt)_5$. [47, 48, 49]

The deposition of Ta_2O_5 films has been achieved by PICVD using either $Ta(OEt)_5$ solutions (30-50 mg/ml) in heptane or tetrahydrofuran, or a new precursor [supplied by Inorg Tech, $Ta(OEt_5)_4(DMAE)$] which has a higher volatility and a considerably lower moisture sensitivity.

Layers were deposited on Si (001) substrates in the temperature range 400-900°C. Dependence of the film growth rate and refractory index on the growth temperature are given in Figure 11.

Using $Ta(OC_2H_5)_5$, the maximum growth rate (11 μm/h) was reached at 650°C. The XRD study revealed that Ta_2O_5 layers deposited at lower temperatures (≤650°C) were amorphous, whereas the layers deposited at the higher temperatures were composed from non-orientated crystallites of the orthorhombic β- Ta_2O_5 phase.

SEM and TEM images of crystalline films deposited at above 700°C revealed a columnar film structure, which leads to a very rough surface and is unsuitable for applications such as capacitors (RMS roughness = 16.5 nm!)

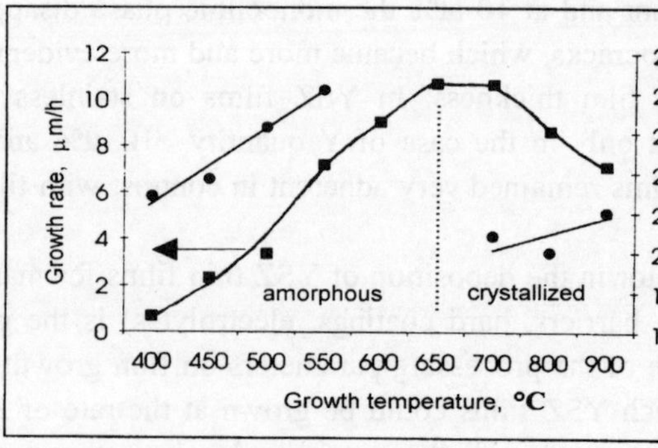

Fig 11. Growth rate and refractive index of Ta_2O_5 as a function of deposition temperature.

The surface of amorphous films was very smooth (RMS roughness = 1.6 nm.). The great difference between the surface roughness of amorphous and crystallised films was clearly visible in the AMF images presented in Figure 12.

The influence of the gas environment during deposition, and post-annealing on the films properties was also studied. The gas environment influenced the carbon content in the films and, after a post-annealing at 850 °C under oxygen (30 minutes) the amorphous films became thoroughly crystallised with a smooth surface. Thermal treatment in fact allowed an increase in the refractive index of up to 2.4.

The use of $Ta(OEt)_4(DMAE)$ as a precursor allows the preparation of Ta_2O_5 at a much lower temperature. With a pressure of 1.3 hPa, the deposition starts at temperatures around 350 °C; the growth rate increases with temperature and reaches a maximum of 25 μm/hour at 500 °C. As with $Ta(OEt)_5$ the oxygen concentration does not influence the growth rate since the precursor contains the amount of oxygen needed to synthesise the oxide. An interesting feature of this precursor is that it may be mixed without chemical interaction with a titanium precursor ($Ti(Oet)_2(DMAE)_2$). It has been reported that the dielectric constant of Ta_2O_5 (from 40 to 120) is greatly enhanced by substitution of TiO_2 [50]. This result has also been observed on sputtered thin films [51], but with a lower amplitude (from 40 to 55). Using a single source containing a mixture of Ti and Ta precursors, we have obtained amorphous and transparent mixed oxide films at 500 °C, with a refractive index of 2.2-2.4. Composition varied between 0 and 30 at% Ti. The samples were observed by SEM, showing a flat surface without precipitates, but the dielectric properties were very poor (due to high current

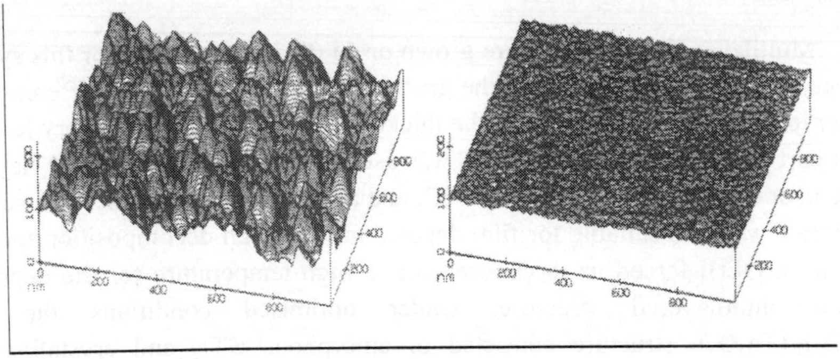

Fig 12. AFM image of a crystallised Ta_2O_5 sample deposited at 700 °C (left), and of an amorphous sample deposited at 450 °C (right). Thicknesses = 350 nm.

leakage). These high leakage currents are probably related to the lack of preparation of the Si surface before deposition: a post-nitridation, and a post-annealing are generally needed to obtain low leakage currents in Ta_2O_5. Further experiments on more suitable substrates are underway.

4.2 Oxide multilayered structures deposited by pulsed injection CVD

One of the most interesting features of PICVD is the simplicity of growing multilayers. Several injectors each connected to the same evaporator are used to inject different solutions alternately. The synthesis of the very complex stackings composed of different layers is of course possible, but our studies have been limited to two components: multilayers and superlattices. All our experiments have been made using two injectors. Among the different structures already grown, we will describe here only examples on crystallised-amorphous or fully amorphous Ta_2O_5/SiO_2, and on epitaxial $YBa_2Cu_3O_{7-x}$ / $PrBa_2Cu_3O_{7-x}$.

4.2.1. $(Ta_2O_5/SiO_2)_n$ multilayers

Multilayered structures consist of alternating thin films of two oxides with very different refractive index, have potential optical applications (e.g. optic guides, filters, anti-reflective layers, supermirrors and glass art). Multilayered structures of Ta_2O_5 (n=2.2-2.4) and SiO_2 (n=1.45) are very suitable for these applications. The achievement of such structures allows the modulation of the refractive index in the depth of material, as well as the production of material for special purposes with the required refractive index.

Multilayered structures were grown on Si (001) substrates. For this synthesis it is very important to co-ordinate the growth conditions of the two different layers in order to ensure precise control of the thickness and growth rate of every layer. In the first part of the experiment $Si(OEt)_4$ was used for the deposition of SiO_2 layers. This precursor starts to decompose at 800°C, and at 850°C it decomposes at a sufficiently high rate which is suitable for film deposition. The high decomposition temperature of the $Si(OEt)_4$ forced us to choose quite a high temperature for the deposition of whole multilayered structure. Under optimised conditions the periodic $(SiO_2)_7/(Ta_2O_5)_7$ structure consisted of amorphous SiO_2 and crystallised Ta_2O_5 layers was deposited at 850°C. (Fig 13)

The growth rates of the two materials were rather similar: 0.33 nm (Ta_2O_5) and 0.2 nm (SiO_2) for each microdose (about 4 mg) of the injected solution. The thickness of Ta_2O_5 and SiO_2 layers in this structure was ~30 nm and ~50 nm respectively. The surface of the multilayered structure was smoother (RMS =8nm.) than in the case of single Ta_2O_5 layers deposited under the same conditions. This is due to the interruption of the columnar growth of Ta_2O_5 by the amorphous SiO_2 layer.

The interfaces in the deposited structure containing crystalline Ta_2O_5 were however too rough for the synthesis of very fine multilayers. In order to obtain smooth interfaces, Ta_2O_5 layers in these structures must be amorphous, so the deposition temperature needed to be decreased and in the next part of the experimental study another Si precursor - $(AcO)_2Si(Ot\text{-}Bu)_2$ (Si(DADBS)) – with a lower decomposition temperature was used.

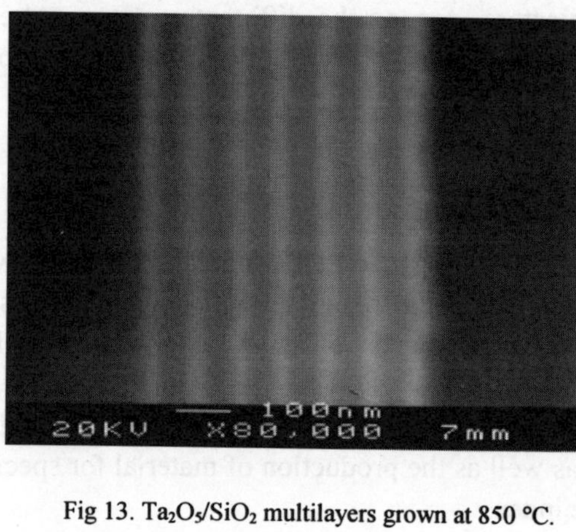

Fig 13. Ta_2O_5/SiO_2 multilayers grown at 850 °C.

A periodic multilayered structure with 25 layers of each oxide, was deposited at 550°C. A cross-section TEM image of such a structure is given in Fig. 14. The average layer thickness was about 5 nm with variation within 10%. The interfaces were distinct, especially for the first layers. A progressive increase in roughness was noted however with the increase in film thickness: this was due to the formation of nodular defects. According to some workers this could be caused by intrinsic mechanical stresses or defects (precipitates) created in the film during the deposition process, nevertheless the completely amorphous heterostructure had distinctly a smoother surface and more distinct interfaces than was the case in structures with crystalline Ta_2O_5. Note that this result is a preliminary experiment and was obtained without any optimisation.

4.2.2. Superlattices ($YBa_2Cu_3O_{7-x}/PrBa_2Cu_3O_{7-x})_n$

The possibilities of depositing regular $(YBa_2Cu_3O_{7-x}/PrBa_2Cu_3O_{7-x})_n$ (YBCO/PrBCO) superstructures using PICVD were investigated. In order precisely to control the thickness of every YBCO and PrBCO layer, the growth rate needed to be decreased compared with that used for YBCO deposition. For this reason the concentration of precursor solutions was divided by six and in this way the thickness of the layer grown by a single microdose was considerably decreased. PrBCO layers were deposited using a solution of Pr, Ba and Cu thd complexes in monoglyme.

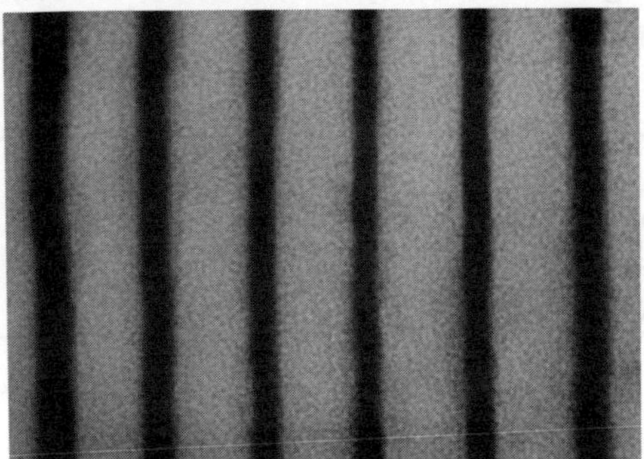

Fig 14. TEM view of an amorphous Ta_2O_5/SiO_2 multilayer grown at 550 °C. The dark lines are Ta_2O_5 (thickness of the black central layer = 3 nm).

After the co-ordination of growth conditions for both compounds, several multilayered structures containing 20 YBCO/PrBCO double layers were deposited. A series of structures were grown by varying the number of injections for YBCO deposition. PrBCO layers were grown at a constant number of injections (100), ensuring that their thickness was constant.

XRD study revealed that the superstructures were epitaxial. The distinct satellite peaks on both sides of 00*l* reflections (Fig. 15) confirmed the regularity of the superstructures. The point at which superlattices were deposited (Λ) was calculated from the position of satellite peaks.

For the synthesis of these samples, the number of injections for the PrBCO layers was fixed at 100, for YBCO, it was varied from 40 to 400. The timing was linearly dependent on the number of injections used for YBCO deposition: thus it was possible to calculate the mean thickness deposited at each injection (0.08 nm/injection for YBCO and 0.04 nm/injection for PrBCO). This allowed the determination of the average thickness of YBCO and PrBCO layers in the different superlattices (Table 2).

The dependence of the critical temperature vs. period of structure was typical for such structures: T_c values decreased with the decreasing of YBCO thickness and this was related to the weakening of the interaction between YBCO layers.

These results showed that PICVD allows the deposition in a very simple way of multilayered nanostructures of high temperature superconductors. The properties of the superlattices grown were in agreement with those obtained by PVD [52,53]. This demonstrated the potential application of this technique in nanotechnology.

Table 2. Properties of a set of YBCO/PrBCO epitaxial superlattices.

Number of injections (YBCO/PrBCO)	Superlattice Period (nm.)	Critical Temperature (K)	ΔTc (K)	Layer thickness (nm) YBCO/PrBCO
40/100	6.8	40	5	2.9/4.6
100/100	11.8	83	1.6	7.3/4.6
160/100	17.3	84	0.8	11.7/4.6
250/100	23.3	89	0.3	18.2/4.6
400/100	33.3	90	0.4	29.2/4.6

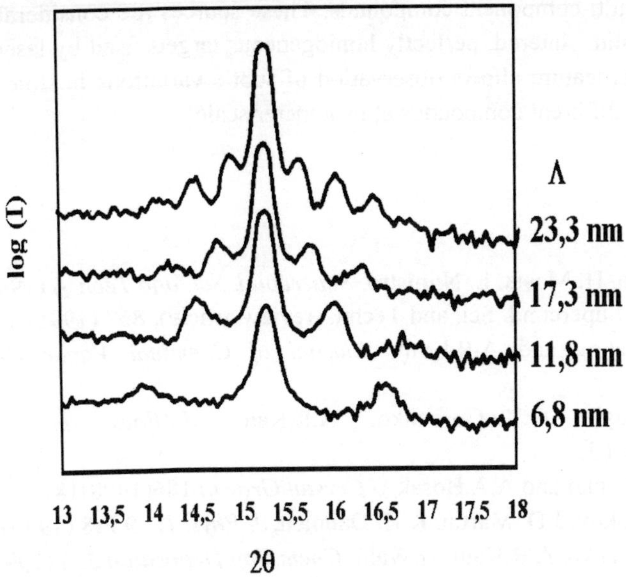

Fig 15. Satellite peaks observed for various YBCO thicknesses in different multilayered samples.

5. Conclusions

For the deposition of complex oxides, the main advantages of the injection technique over normal CVD sources are:

- a strong enhancement of the reproducibility of the properties of the layers.
- very simple and versatile control of the growth rate, with two computer-controlled parameters (frequency and electrical pulse width)

The use of a single source for the synthesis of multicomponent compounds simplifies this process considerably in comparison with classical bubblers or sublimators. The solution injected is flash-volatilised, and the time of transfer between the evaporator and the substrate is very short (<< 1 sec.): even unstable precursors may be used with a good reproducibility of the vapour pressure. Complex compounds containing Ba, Sr or rare earth, which are very difficult to obtain by classical CVD process, are easily and reproducibly synthesised by injection. The main advantage over PVD techniques is the flexibility of the system, which is related to the simplicity of the source being only a weighted mixture of liquids, injected using a low cost, highly reliable system. This system uses a liquid source, consisting simply of a weighted solution, or "cocktail", of precursors for the synthesis of multi-component compounds. These sources are considerably easier to use than the solid, sintered, perfectly homogeneous targets used by laser ablation or sputtering. This feature allows observation of subtle variations in stoichiometry, or the stacking of different compounds at nanometer scale.

References

1. M. Leskela, H. Mosla, L. Niinistro, *Supercond. Sci. and Tech.* **6** (1993) 627
2. M. Jergel, Supercond. Sci. and Techn. (review article), **867** (1995) 12
3. L.S.Chuprakov and A.R.Kaul, *Journal of Chemical Vapor Deposition* **2** (1993)123
4. A.A.Molodyk, O.Yu.Gorbenko, A.R.Kaul, *J.Alloys and Compounds* **951**(1997)303.
5. O.Yu.Gorbenko and A.A.Bosak, *J.Crystal Growth* **186**(1998)181
6. I.S. Chuprakov, J.D. Martin, K.H. Dahmen, *J. Phys. IV*, **9** Pr8 (1999) 901
7. O.Yu.Gorbenko, A.R.Kaul, G.Wahl, *Chem.Vap.Deposition* **3,** 4 (1997)
8. L. Klippe, R. Stolle, W. Decker, A. Nurnberg, G. Wahl, Yu Gorbenko YU Erhokin, I. Graboy, A.R. Kaul, S.A. Pozigun, *J. of Alloys and Compounds* **251** (1997) 264
9. V.N. Fuflyigin, A.R.Kaul and S.Pozigun, *J.Mater. Sci.***30** (1995)4431
10. O.Yu.Gorbenko, A.R.Kaul, A.A.Bosak, I.E.Graboy, H.W.Zandbergen, V.L.Svetchnikov, N.A.Babushkina, L.M.Belova, K.I.Kugel, Solid St. Comm **114** . (2000) 407.
11. D.B. Studebaker,J. Zhang, T.J. Marks, Y.Y. wang, V.P. david, J.L. Schindler, Appl. Phys. Lett. 72, **10** (1998) 1253
12. S. Matsuno, T. Unemura, F. Uchikawa and B. Ikeda, *Jpn. J. Appl. Phys.,* **34** (1995) 2293
13. J.P. Sénateur, R. Madar, O. Thomas, F. Weiss, A. Abrutis Patent n° 93/08838 - Licensed J.I.P. Elec - 11, chemin du Vieux Chène - 38240 MEYLAN - France – Fax 33 4 76 04 81 40
14. J.P. Sénateur, A. Abrutis, F. Felten, F. Weiss, O. Thomas, R. Madar. Advances in Inorganic Films and Coatings. Ed. P.Vincenzini, Techna srl.(Italy), 1995, p. 161-166.

15. F. Weiss, J.P. Sénateur, J. Lindner, V. Galindo, C. Dubourdieu, A. Abrutis J. Phys. IV **9** vol. 1 Pr8-283 (1999)
16. F. Felten, J.P. Senateur, F. Weiss, R. Madar, A. Abrutis. *J. Physique IV*, **C5** (1995) 1079-1086.
17. F. Felten, J.P. Sénateur, M. Labeau, K. Yu-Zhang, A. Abrutis. Thin Solid Films. **296** (1997) 79-81.
18. C. Jimenez, M. Paillous, R. Madar, J.P. Sénateur, A.C. Jones J.Phys.IV **9** Vol 2 pr8 -569 (1999)
19. Bourhila N., F. Felten, J.P. Sénateur, Schuster F., Madar R., A. Abrutis. Electrochem. Society Proceedings, **25** (1997) 417-424.
20. A. Abrutis, V. Kubilius, V. Bigelytë, A. Teiserskis, Z. Saltytë, J.P. Sénateur, F. Weiss. Material Letters, **31** (1997) 201-207.
21. A. Abrutis, J.P. Sénateur, F. Weiss, V. Bigelytë, A. Teiserskis, V. Kubilius, V. Galindo, S. Balevièius. J. Crystal Growth, **191** (1998) 79-83.
22. F. Weiss, U. Schmatz, A. Pisch, F. Felten, S. Pignard, J.P. Sénateur, A. Abrutis, K. Frohlich, D. Selbman, L. Klippe. J. Alloys Compounds, **251** (1997) 264-269
23. J.P. Sénateur, F. Felten, S. Pignard, F. Weiss, A. Abrutis, V. Bigelytë, A. Teiserskis, Z. Saltyë, B. Vengalis. J. Alloys Compounds, **251** (1997) 288-291
24. A. Abrutis, J.P. Sénateur, F. Weiss, V. Kubilius, V. Bigelytë, Z. Saltyë, B. Vengalis, A. Jukna. Superconductor Science Technology. **10** (1997) 959-965.
25. A. Abrutis, V. Kubilius, A. Teiserskis, V. Bigelytë, V. Galindo, F. Weiss, J.P. Sénateur, B. Vengalis, A. Jukna, R. Butkutë. Thin Solid Films, **311** (1997) 251-258.
26. V. Galindo, J.P. Sénateur, A. Abrutis, A. Teiserskis, F. Weiss - J. Cryst. growth **208** (2000) P. 357-364
27. J. Lindner, F. Weiss, J.P. Sénateur ,W. Haessler, G. Gobernik, S. Oswald, J. Santiso, A. Figueras - J. Phys. IV **9** pr8 (1999) 283
28. S. Pignard, J.P. Sénateur, H. Vincent, Kreisel J., A. Abrutis. J. Physique IV, **7** C1- (1997) 483-484
29. S. Pignard, H. Vincent, J.P. Sénateur Thin Solid Films, **350**, (1999) 119
30. S. Pignard, H. Vincent, J.P. Sénateur, J. Pierre - J. of Alloys and Compounds, **262**, (1997) 157
31. S. Pignard, H. Vincent, J.P. Sénateur, P.H. Giauque - Thin Solid films **347** (1999) 161

32. S. Pignard, H. Vincent, J.P. Sénateur, J. Pierre, A. Abrutis. J. Applied Physics. **82** (1997) 4445-4448.
33. C. Dubourdieu, M. Audier, J.P. Sénateur, J. Pierre - J. App. Phys. **86** n° 12 (1999) 6945
34. S. Pignard, H. Vincent, J.P. Sénateur, K. Fröhlich, J. Souc Appl. Phys. Lett., **73** (7 (1998) 999
35. A.A. Bosak, O. Yu. Gorbenko, A.R. Kaul, I.E. Graboy, C. Dubourdieu, J.P. Sénateur, H.W. Zandbergen - JMMM **211** (2000) p. 61-66
36. A.Abrutis, V.Plausinaitienë, A. Teiserskis, V. Kubilius, Z .Saltytë, J.P. Sénateur, L. Dapkus.. J.Phys.IV **9** pr8 p. 2 (1999)
37. A.Abrutis. J.Phys.IV, **9** pr8-683 (1999)
38. Molodyk A.A., Novozhilov M.A., Graboy I.E., Gorbenko O.Yu., Korsakov I.E., Kaul A.R, *Electrochem.Soc.Proc.*,**97**-25, (1997) 1152.
39. M.A.Novozhilov, A.R.Kaul, O.Yu.Gorbenko,I.E.Graboy, G.Wahl and U.Krause, *J.Phys.IV France* **9** (1999) 629
40. G. Wahl, O. Stadel, L. Klippe, A.R. Kaul, O.Y. Gorbenko, S.V. samoylenkov *ProcedingsNato Advanced Workshop HTSC Y,* Moscow March 24-29 (1998)Ed by Van Tanderloo – Nato Sciences Series 3 high technologies **62** (1999) 79
41. B.H. Moecly, S.E. Russek, D.K. Lathrop, R.A. burman, J. Li, J.W. Maye, *Appl. Phys Lett.* **57** (1990) 1687
42. K.H. Young, McD. Robinson, G.V. Negrete, J. Zhao, C.S. Chern, Y.Q. Li, P. Norris, *Appl. Phys. Lett.,* **61** (1992) 543
43. P.C. Chou, Q. Zhong, Q.L. Li, A. Ignatiev, C.Y. Wang, E.E. Deal, J.G. Chen, *Mater. Res. Soc. Symp. Proc.* **335** (1994) 279
44. T. Clausen, J.L. Skov, C.S. Jacobsen, K.R. Bukh, M.V. Bollinger, B.P. Tobiasen, M.P. Sager, I. Chorkendorff, J. Larsen, *J. Appl. Phys.,* **79** (1996) 7062.
45. J. Hudner, M. Östling, H. Ohlsen, L. Stolt, P. Nordblad, M. Ottoson, J.C. Villegier,, H. Moriceau, F. Weiss, O. Thomas,*J. Appl. Phys.,* **73** (1993) 3096.
46. J.P. Hong, J.S. Lee, *Appl. Phys, Lett.,* **68** (1996) 3034.
47. J.L. Autran, P. Paillet, J.L. Leray, R.A.B. Devine *Sensors and Actuators A-51 (1995) 5*
48. W. Kern, A. Chen, N. Sandler J. de Phys. IV. C2 Vol. 1 (1991)
49. Xu Li, M. L. Hitchman, S.H. Shamlian Electrochemical Soc. Proceedings Vol 97-25; Proceedings Euro CVD 11 Paris (1997) 1246
50. R.F. Cava, W.F. Peck, J.J. Krajewski *Letters to Nature* **337** (1995) 215

51. J.Y. Gan, Y.C. Chan, T.B. Wu, *Appl. Phys. Lett.* ***72*** 3 (1998) 332
52. J.M. Triscone, O. Fischer, O. Brunner, L. Antognazza, A.D. Kent, M.G. Karkut, *Phys. Rev. Lett.* **64** (1990) 804
53. J.P. Contour, S. Sant, D. Ravesolosona, B. Fischer, L. Partagan, *Jpn. J. Appl. Phys.* **32** (1993) L1134

CHAPTER FOUR

NOVEL SYNTHESIS OF NANOCRYSTALLINE DIAMOND FILMS

Stanislaw Mitura

Thin Films Division, Institute of Materials Science, Technical University of Lodz
90-924 Lodz, ul. Stefanowskiego 1/15, POLAND; E-mail: mitura@p.lodz.pl

1. Introduction

Without the element carbon, life as we know would not exist. Carbon provides the framework for all tissues of plants and animals. These tissues are built of elements grouped in chains and rings made of carbon atoms. Carbon has been named "black gold" which shows a limited imagination of linguists because how such an element as carbon with absolute, unrestricted possibilities could be compared with such a common metal as gold is inexplicable. Among more than 15 million known chemical compounds a major proportion consists of carbon compounds. Starting from the most simple products of metabolism like carbide through multimolecular polymers or proteins - carbon is a fundamental adhesive and a guarantee of structure stability.

Such an interesting element has obviously been investigated and because chemistry is based on outer-shell electrons (also known as valence electrons), it is believed that the secret of carbon is hidden in these valence electrons. From the Periodic Table one can see that carbon belongs to the IVth group and 2nd period. Out of six electrons surrounding the carbon atom nuclei, two are able to create chemical bonds. When a certain amount of energy is used, the carbon atom is in the excited state and then four outer-shell electrons determine the variety of carbon structure both in chemical compounds and in pure carbon modifications.

We all remember that our school books contained information about two main forms of carbon: diamond and graphite. In diamond lattice, each carbon atom is connected to four others; all valence electrons are involved in creating chemical bonds which are very strong and difficult to break. The diamond crystal is an example of the valence structure, because each atom of carbon, situated in a tetrahedron, is connected with another by a covalent bond. These bonds are said to be very stable with quite a high melting point which is why diamond is considered to be one of the hardest material. Besides it does not conduct electricity, but it is an excellent heat conductor, although it does not conduct heat in the same way as metals (by electron drifting). Its good capacity for heat conduction comes from its structure (that is shown in Fig. 1).

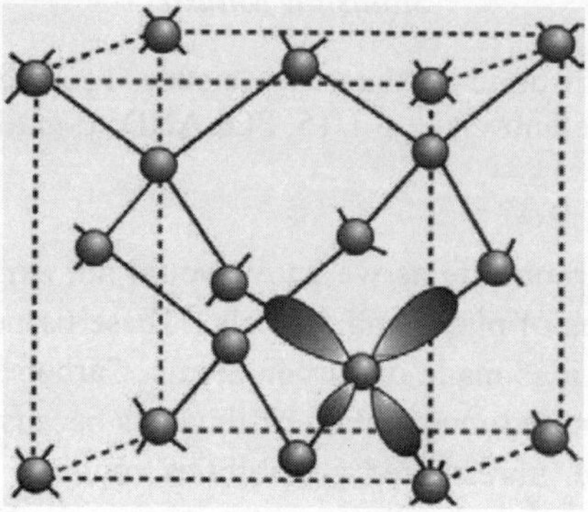

Figure 1. Diamond structure with σsp^3 bonds around one of carbon atoms.

The second modification, graphite has quite the reverse characteristics: it splits easily and conducts electricity. Graphite is an example of an intermediate structure between valence and molecular crystals. The graphite lattice consists of parallel layers of atoms in flat sheets, placed in honeycomb form, in which carbon atoms lie in the corners of the hexagons. Each atom is joined to its three neighbours by the covalent bonds and there are only small interactions (van der Walls forces) between layers. This type of molecular bond does not require a great deal of energy to be destroyed. This type of molecular bond does not require a great deal of energy to be destroyed. In fact graphite occurs not only in a hexagonal modification, but also in a cubic or rombohedron shape as a result of displacements of particular sheets or layers. Because each carbon atom is bound to the three closest atoms, valence electrons are not fully involved and electric conduction along sheets may occur. Fig. 2 illustrates the graphite structure.

All these are facts which have been known for a long time in the history of chemistry. In fact carbon as an element is already well researched and tested but nevertheless it is necessary to bring the knowledge up to date as new carbon forms have been discovered and methods of fabrication are being described. Apart from diamond and graphite there are other carbon modifications: carbines, fullerens and nanotubes.

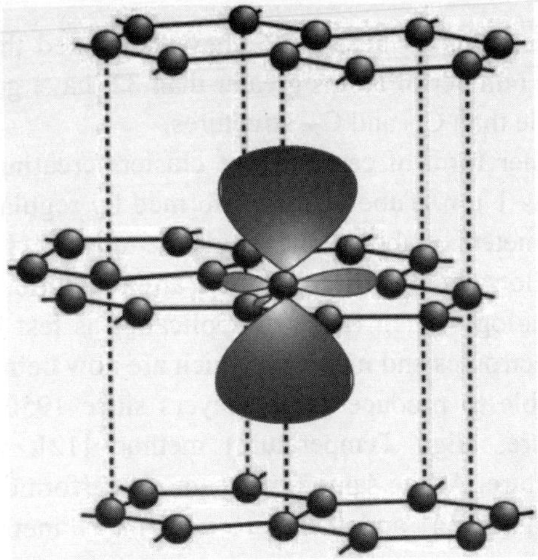

Figure 2. Graphite structure with σsp^2 Πp bonds around one of carbon atoms.

Carbines are those about which the least is known: they are an allotropic form of carbon. The earliest information published about them was in the sixties in the work of Gorsey and Donney [1], who described discovery of a new allotropic form of carbon, called chaoite or carbon from the Ries crater [2,3] (the name comes from the crater where this carbon form was found). According to the authors' hypothesis, chaoite was formed from graphite when a meteorite was subjected to high temperatures and pressures just before collision. At the same time Kasatochkin and his co-workers [4,5] were carrying out theoretical and experimental works concerning the synthesis of a so-called carbon polymer which was later called carbine. They hold the thesis that both diamond and graphite are polymers, but diamond is an impregnated polymer belonging to the group of alkans, and graphite is an impregnated polymer belonging to the aromatic group. There are now more than six carbine modifications, but carbines α and β are the best investigated [6,7].

Fullerens [8,9] form another modification of pure carbon. Their archetype is the molecule C_{60}, probably the most circular one. This structure was first studied by American engineer and philosopher R. Buckminster Fuller [10] and the new molecule was thus named buckminsterfullerene. C_{60} has good stability derived from its structure which consists of equilateral hexagons and pentagons. In the XVIII century the Swiss mathematician Leonhard Euler proved that each closed sphere should consist of 16 pentagons while the number of hexagons is arbitrary. Robert F. Curl and Robert E. Smalley from Rice University have investigated carbon clusters [9,10] and they suggest the arrangement of atoms in the C_{70} cluster which gives rise to the fact that C_{70} contains 25 hexagons. They also stated that all carbon clusters consisting of an even number of atoms greater than 32, have good stability although they are still less stable than C_{60} and C_{70} structures.

Nanotubes - another form of carbon - are clusters creating tubes whose length may reach as much as 1 μm. Tube walls are formed by regular hexagons and there are channels with diameters of about 2 nm inside the tubes [11].

All mentioned allotropic forms of carbon are revolutionary and change our current view of technology. Their range of application is fast increasing, especially for applications in electronics and medicine which are now being considered.

It has been possible to produce carbon layers since 1950 but this was by the HPHT (High Pressure, High Temperature) method [12], which required high pressure and temperature. At the same time methods performed under low pressure were being described [13,14] and there are now many methods of carbon layer synthesis [15].

At the moment the most popular method for polycrystalline diamond (PCD) synthesis is the Microwave Plasma Chemical Vapour Deposition (MW PCVD). In the MW PCVD method, Kamo, Sato, Matsumoto and Setaka [16] were the first to report that carbon-containing gases and hydrogen molecules were dissociated by microwave radiation at a frequency of 2.45 GHz. The same phenomenon was found in a domestic microwave oven.

Another popular method is Hot Filament Chemical Vapour Deposition CVD [15]. The activation of the gas phase by thermally-emitted electrons from the hot filament seems in our view to be the simplest and most economical method on a laboratory scale: by this method a mixture of about 1% methane in hydrogen enters the reactor and diamond films are formed on a substrate heated up to ~900°C and mounted a few millimetres beneath the hot filament. A variation of this method uses additional radio frequency plasma between the filament and the negatively self-biased substrate holder (HF RFPCVD) [17,18].

Boudina, Fitzer and Wahl reported very high growth rates of polycrystalline diamond layers of 300 to 500 micrometers per hour [19]. They manufactured PCD layers by the Arc-Discharge Plasma-Jet CVD method.

2. Experimental

Bubenzer, Dischler and Nyaiesh [20] developed a method of hydrocarbon decomposition in a radio frequency electric field, depositing hydrogenated amorphous carbon (a-C:H) on to substrates placed on a negative self-biased electrode in the reactor chamber under reduced pressure. According to the authors, changing the negative self-biased potential of the RF electrode and hydrocarbon pressure in the reactor, can affect the temperature of the substrate and the ion energy. These are the parameters that have a direct effect on such properties of the layers as energy gap, optical absorption and refractive index. The negative self-biased potential V_e, which is a universal parameter characterising layer properties, is directly proportional to electron temperature and, taking into account the plasma potential, it has an effect on the average energy of the ions which collide with the RF electrode. The parameter V_e determines the total energy transferred from plasma to the layer during deposition [21,22].

Amorphous carbon layers consist of carbon atoms differing in the hybridisation of their electron orbitals. Dischler, Bubenzer and Koidl [21] determined the ratio of atoms with hybridisation of electron orbitals of type sp^3, sp^2 and sp^1 in hard carbon layers at 68:30:2.

Enke, Dimigen and Hubsch [23] determined precisely the range of the soft and hard layers. Enke called them ion carbon, (i-C), to determine the predominant role of ions in the process of formation.

Vora and Moravec [24] modified Holland and Ojha's [25,26] reactor by additional negative bias of R.F. electrode. The carbon layers which they obtained consist of an amorphous warp with in-built diamond crystals. Has, Mitura and Wendler [27] confirmed the presence of diamond microcrystals embedded in an amorphous film.

2.1. Nucleation of diamond particles

The aim of the study was to determine the conditions and to explain the mechanism of the formation of crystalline inclusions which are present in the amorphous warp of carbon films; these are shown built into the layer in Fig. 3. The layers were deposited in a plasmachemical reactor as previously described [22,28-31].

Carbon films formed by hydrocarbon decomposition in the RF electric field are characterised by an amorphous structure with a few crystalline inclusions. The layer shown in Fig. 3 has an extremely large number of crystallites which can be seen in microphotography as dark points in a bright field. In the dark field (Fig. 3(c)), surrounding diamond crystallites, regions brighter than the rest of the layer can be seen: these are the nuclei of diamond crystallisation.

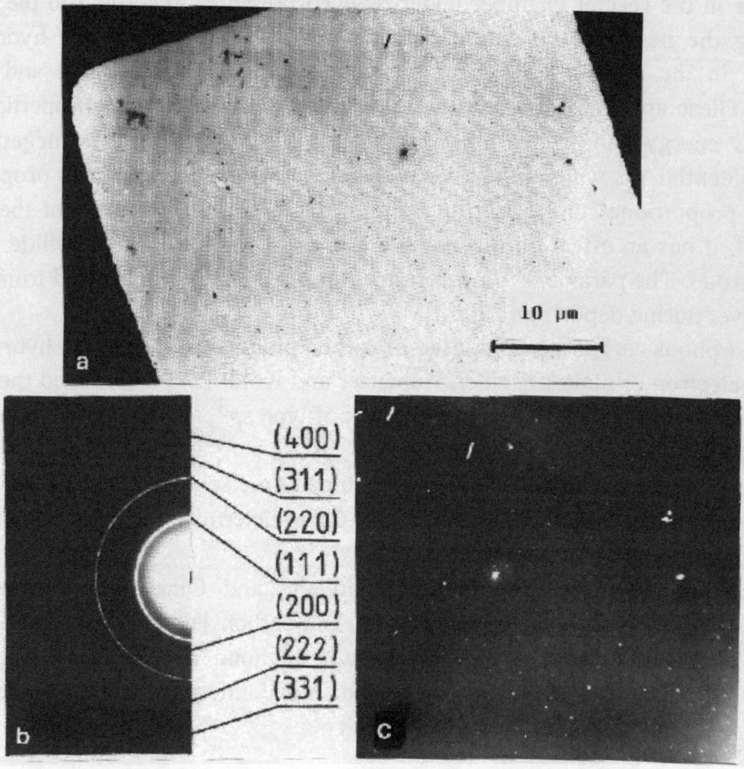

Figure 3. Carbon film obtained by hydrocarbon decomposition in the RF electric field (RF CVD), Philips TEM 300, (a) image in the bright field, (b) diffraction pattern (diamond), (c) image in the dark field.

Apart from single microcrystals of diamond, shown in Figs. 3(a) and 1(b), there are aggregates of microcrystals in the layers presented in a micrograph in Fig. 4.

At the time when amorphous carbon layers with crystalline inclusions are formed on the negative self-biased R.F. electrode (Figs. 3 and 4), polymer films are deposited on the other grounded electrode which consist of an amorphous warp and polymer powders.

The size of the diamond inclusions is similar to that of a powdery polymer, embedded in the polymer film formed on the grounded electrode. Thompson and Mayhan [32] reported that the size of polymer powders decreased with an increase of the R.F. glow discharge power and that their density increased with the increase in discharge power. Thompson and Mayhan [32] proved that polymer powders were formed in plasma and were positively charged.

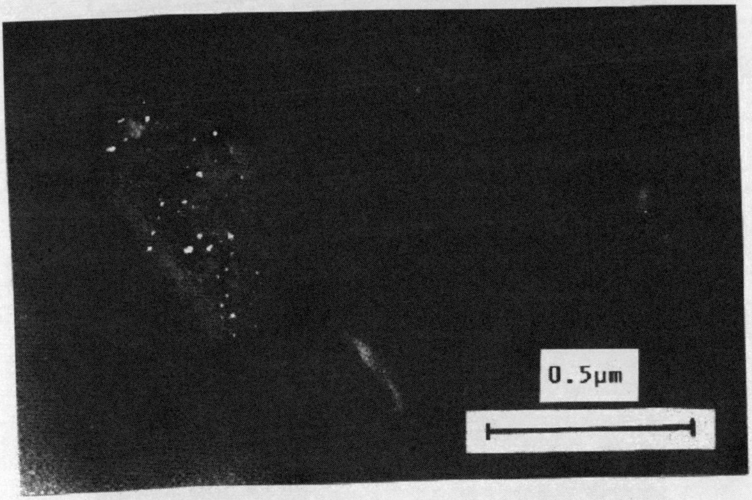

Figure 4. Microscopic image of a group of diamond crystals embedded in the amorphous carbon layer; dark field - electrons reflected from (220).

Grebowicz, Pakula, Wrobel and Kryszewski [33] proved that a powdery polymer was formed as a result of a polymerisation in the gas phase.

The author of the present paper [22] assumed that the first stage of the formation of crystalline inclusions was analogous to the process of volumetric polymerisation: the difference is in applying different power to the plasma by collision of electrons and gas molecules.

A microscopic image of carbon film with powder particles deposited on to its surface is shown in Fig. 5. Spherical powders form spatial agglomerates which lie only on the surface of a carbon layer and often are not bound to it. This can be seen clearly in Fig. 5. Powders consist of aggregates of diamond microcrystals, which are shown in an electron diffraction pattern in Fig. 5b. Spatial powder conglomerates form "bridges" over destroyed fragments of the carbon layer; such characteristic powder conglomerates unbound to the layer are seen on the left and in the centre of the picture, in the bright field (Fig. 5c).

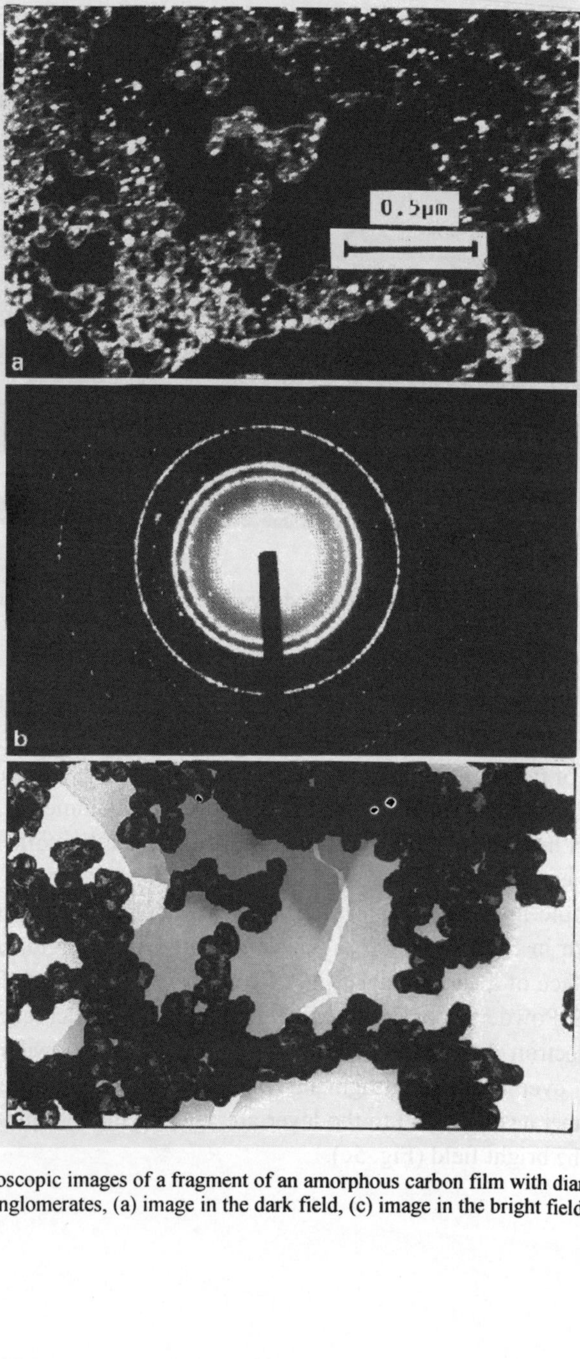

Figure 5. Microscopic images of a fragment of an amorphous carbon film with diamond powder conglomerates, (a) image in the dark field, (c) image in the bright field.

Diamond powder particles (DPP) which are deposited on to the carbon layer during the process can be easily removed from the surface by a brush. This causes neither destruction nor weakening of the layer. Subsequent stages of an inclusion of the DPP agglomerates in the carbon layer formed on an RF electrode were also observed [34]. In microphotographs made in both bright and dark fields subsequent stages of the layer formation were recorded by means of an electron microscope. In this way an amorphous carbon layer with crystalline inclusions is formed and the structure of these inclusions depends on the value of the negative potential of the self-biased RF electrode [22,34].

A separate problem in the RF CVD process is the utilisation of additional electrons introduced into the region in which plasma-chemical reactions take place. It was shown previously [17,18] that the application of additional hot-wire electrons allowed the formation of polycrystalline diamond films. For this purpose a hot filament of a strictly determined potential should be used which allows electrons of specific energy to be introduced into the plasma [18,22].

Before the analysis of the results the authors determined basic parameters with reference to the RF plasma. A characteristic feature of the RF plasma is a lack of thermodynamic equilibrium which is expressed by differences in kinetic energy of electrons, ions and inert gas particles. The measure of the kinetic energy of plasma components is temperature-defined by the equation: $E_k = 3/2\, k\, T$, where: E_k is the kinetic energy of a particle and k is the Boltzmann constant equal $1.38 \times 10^{-23}\ JK^{-1}$.

When describing the RF plasma using temperatures of its components, the following inequality is derived [22]: $T_e > T_i > T_g$, where T_e, T_i, T_g are temperatures of electrons, ions and neutral gas components respectively.

In plasma excited by the RF electric field higher than 1 MHz the mobility of electrons is several orders higher than that of ions. Ions are practically immobile; the RF field does not affect them. The entire electric energy of the RF field is transmitted to the plasma region by collisions of electrons with ions and inert gas particles. Plasma-chemical processes that take place in the RF electric field occur only by collisions of electrons with other gas components.

The result of the great difference in mobility of electrons and ions in the RF plasma is called the Debye potential. It consists of a negative self-bias of all isolated bodies embedded in the plasma to the potential [22]:

$$V = \frac{kT_e}{2e} \ln \frac{T_e\, m_i}{T_i\, m_e} \qquad (1)$$

where V is the negative potential of the self-bias of isolated bodies embedded in the plasma, T_e and T_i are temperatures of electrons and ions respectively; m_e and m_j are masses of electrons and ions respectively, and k is the Boltzmann constant.

Since the ion temperature T_i is low compared with the temperature of electrons T_e, in practice (as in the case of RF plasma) the Debye potential is a function of the electron temperature. The potential determined by equation (1) corresponds to an ideal state - it refers to a very small sample placed in the RF plasma of infinite volume. When the plasma volume in the chamber is finite, the importance of grounding the reactor chamber walls is greater. The potential of self-biased bodies isolated from plasma reaches a maximum value equal to that given by eq. (1) when the surface of a body embedded in the plasma tends to zero [35,36]. The ratio of sample surface area to the surface area of grounded walls of the plasma reactor determine this potential. This quotient with the specific coefficient according to the reactor structure is constant for a reactor of determined geometry. In eq. (18) an additional constant will also appear whose value occurs in the interval from 1 to 4 [22,35,36]. With the assumption that ion temperature changes slightly with the change of the RF energy supplied by the generator, the self-bias potential appears to be a function of the electron temperature, T_e:

$$V = f(T_e) \tag{2}$$

When the phenomena discussed above are considered for excitation of plasma in the RF electric field in a diode plasma reactor, the electrode to which the RF energy is supplied through a capacitor is negatively self-biased. The value of the self-biased potential is proportional to the electron temperature T_e.

The authors state that the process of the diamond formation under thermodynamic instability, in the presence of plasma excited in the RF electric field, consists of the following stages: (a) critical nuclei are formed on ions in the gas phase; (b) an amorphous carbon film is formed from these nuclei and carbon ions reaching the negatively self-biased RF electrode. This carbon film contains a fraction of carbon atoms with an sp^3 type of hybridization of electron orbitals, whereas in the activation of growth of the diamond nuclei in the gas phase by collisions with electrons, a polycrystalline diamond film is formed.

As a result of collisions between electrons and carbon atoms in a growing nucleus, one of many electron configurations typical for carbon is formed and the final result is a formation of microcrystals in the gas phase. These observations proved that the structure of these microcrystals is dependent on the negative self-bias potential of the RF electrode, that is on the electron temperature, i.e. the kinetic energy of electrons in the plasma. There is a critical range of the electron temperature of plasma, i.e. the kinetic energy of electrons at which diamond microcrystals are formed. In another range microcrystals of carbines, graphite or powders composed of an amorphous polymer are formed.

2.2. Synthesis of Nanocrystalline Diamond Coatings

In this chapter the results of experimental studies concerning the manufacture of amorphous carbon coatings obtained by a new method of the RF dense plasma CVD are presented. The substrate was coated with superfine crystalline diamond, deposited by the dense plasma method. The basis of this method [29] is to excite plasma in methane with nitrogen, or other hydrocarbons with nitrogen, in an RF electric field at a relatively high gas pressure of about 20 to 400 Pa.

A schematic diagram of the apparatus is shown in Figure 6. The substrates were placed on the RF-powered negatively self-biased electrode.

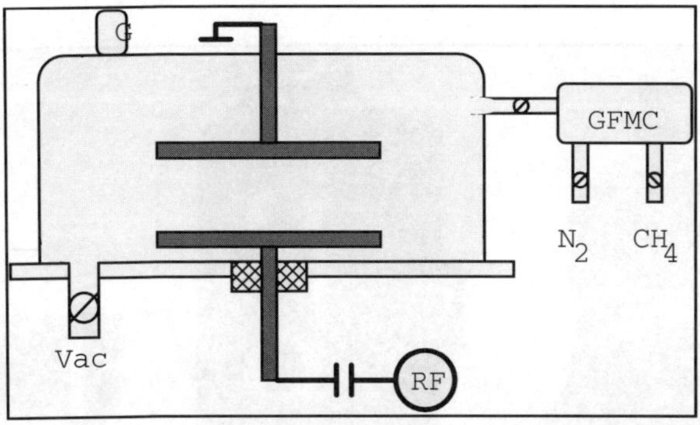

Figure 6. The apparatus for carbon film synthesis by the RF decomposition of methane. RF, 27.12 MHz generator; GFMC, gas feeder with a microcomputer controller; G, vacuum gauge; Vac, vacuum unit.

Experimental details for investigations are given in Table 1. A photo of the plasma-chemical reactor chamber is show in Fig. 7.

Carbon films are somewhat amorphous and it should be emphasised that they are chemically homogeneous. Ellipsometric measurements of the refractive index for DLC and nanocrystalline diamond films (NCD) highlight a significant difference: DLC films are characterised by the refractive index n = 1.9. The average value for the refractive index in the present NCD carbon films n = 2.4 was obtained.

Table 1. Main parameters and growth conditions for the RF dense methane plasma CVD process.

Substrates	AISI 316L or Ti
Gas	Methane with Nitrogen
Pressure	20 to 400 Pa
RF generator	27.12 MHz, 2 kW
Negative self-bias voltage of RF-powered electrode	400 to 1200V
Electrode-to-electrode distance	50 to 330 mm

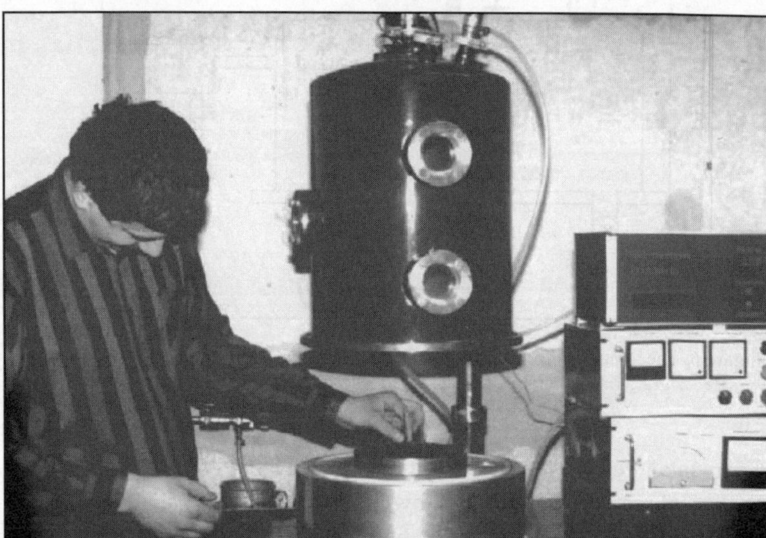

Figure 7. Plasma-chemical reactor chamber with exposition of medical screw.

3. Characterisation

The specimens obtained have been tested to reveal their structure and their biological, mechanical and chemical resistance. The following investigations and tests have been undertaken.

3.1. Structural investigations

A TEM microscopic image of carbon film deposited on to a metal surface shows very small diamond crystallites [17] of grain size similar to that from an atomic force microscopy (AFM) image [37] (compare Figure 8).

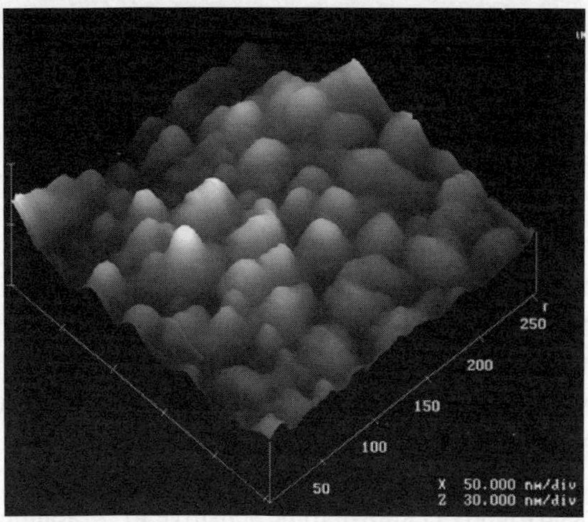

Figure 8. AFM image of NCD film deposited by RF dense PCVD method.

In this study, atomic force microscopy (AFM) is used to examine the morphology of the surface of carbon coatings obtained by RF PCVD deposition processes [37]. AFM seems to be one of the most suitable techniques for investigating the nucleation and growth of films on substrates, although atomic resolution is almost impossible to achieve. This technique allows identification of nuclei which are too small to be recognised with their characteristic cube-octahedral shape of diamond crystallites by scanning electron microscopy (SEM). Figure 8 shows the surface of an NCD layer and Figure 9 shows the surface of a DLC film for comparison.

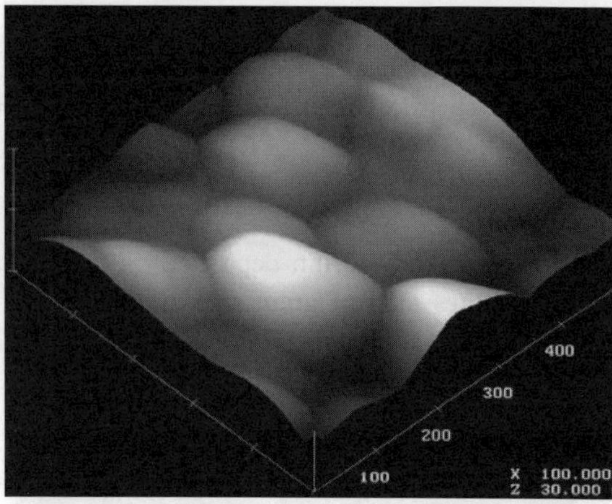

Figure 9. AFM image of DLC film deposited by RF PCVD method.

3.2. Characterisation of the Raman spectra

Additional structural investigations of the carbon DLC and NCD films were carried out by micro-Raman spectroscopy. The Raman spectroscopy of different forms of carbon has been studied extensively [38-42]. The Raman spectrum of an NCD film (see Figure 10) shows a sharp peak at about 1340 cm^{-1} which is usually referred to a diamond peak; the second peak emerges at 1600 cm^{-1} and is referred to as a graphite peak.

The Raman spectroscopy is known to be sensitive to "graphitic" defects [38,42], and the literature reports that the relative scattering efficiencies of diamond and graphite differ by as much as a factor of 50, which has been attributed to resonance effects in graphite. Given the relative sensitivities of the Raman technique to diamond and graphite at the wavelength used, it can be estimated that the film contains a low percentage of graphite. The full width at half-maximum of the 1332 peak can be attributed to the very small grain size of diamond nanocrystals. There is no evidence of a strictly amorphous phase. This is a good quality diamond nanocrystalline film.

Positions of all the peaks existing in this spectra are included in Table 2.

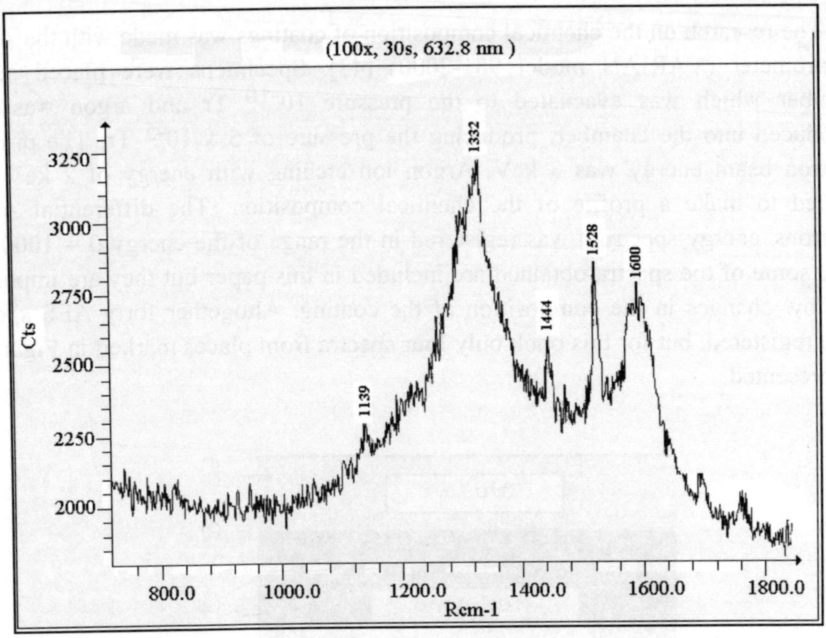

Figure 10. The Raman spectrum of an NCD film.

Table 2. The positions of the Raman peaks.

Peak cm^{-1}]	Phase	Bond
1139	Nanocrystalline diamond	σsp^3
1332	Diamond	σsp^3
1444	Nanocrystalline diamond	σsp^3
1528	Carbyne	sp^1
1600	Nanocrystalline graphite	σsp^2

3.3. AES and microX-ray investigations of substrate/film interface

A specificity of the layers deposited by the dense plasma method, which have very good adhesion to steel, is explained by the authors on the basis of investigations carried out by Auger electron spectroscopy (AES) and microX-Ray Spectroscopy.

The research on the chemical composition of coatings was made with the Auger spectrometer (VARIAN model 981-2000) [43]. Specimens were placed in the chamber which was evacuated to the pressure 10^{-10} Tr and argon was then introduced into the chamber, producing the pressure of 5 x 10^{-5} Tr. The primary electron beam energy was 3 keV. Argon ion etching with energy of 2 keV was applied to make a profile of the chemical composition. The differential Auger electrons' energy spectrum was registered in the range of the energy 0 ÷ 1000 eV. Only some of the spectra obtained are included in this paper but they are important to show changes in the composition of the coating. Altogether forty AES spectra were registered, but for this book only four spectra from places marked in Figure 11 are presented.

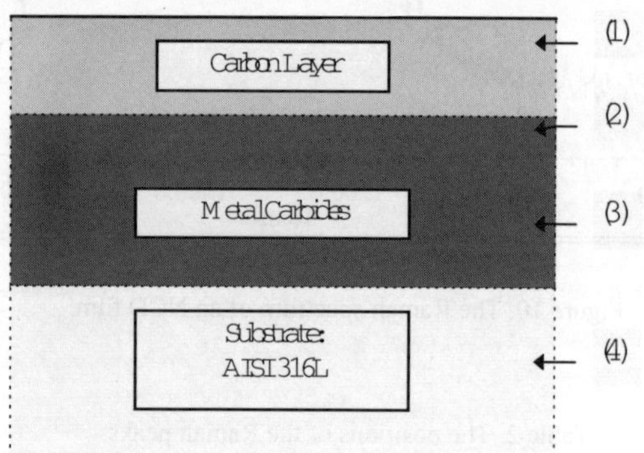

Figure 11. The diagrammatic model of the coating with marked places of AES analysis shown in Figure 12.

The results of AES studies are presented graphically in Figure 12. N° 1 in this figure stands for analysis made directly from the film surface without ion etching. The surface is clean and does not contain any elements apart from adsorbed oxygen and nitrogen. N° 2 shows an analysis made after 13 minutes of ion etching by a process consisting of several stages. In the spectrum, apart from the very large carbon peak, an Auger 215 eV argon peak was found, which is the result of argon ion etching. This effect can be found in all the following analyses. Small iron peaks were also found, which may be connected with plasma chemical processes in the gas phase and 161 eV, 186 eV, and 221 eV molybdenum and 447 eV, 459 eV, 489 eV, 529 eV, and 571 eV chromium peaks were also noted.

N° 3 shows the analysis made after the next ion etching process (15 minutes). The KLL 272 eV carbon peak, especially in its low energy section has a typical metal carbide shape. Auger peaks for Mo and, in particular, for Fe and Cr became stronger and 775-783 eV and 848 eV peaks for nickel were also observed.

N° 4 shows the analysis made after removing the carbon coating from the substrate (AISI 316L steel) by an ion etching process (about 30 minutes). There was almost no carbon in this spectrum. The analysis of the changes in the shape of the C KLL 272 eV carbon peak [44] and the distribution of elements being noted as a source of carbides, indicate the possibility of creating an interface rich in carbides between the carbon film and the steel substrate.

Investigations of the concentrations of elements (C, Cr, Mo, Ni, Mn) were carried out on the scanning electron microscope (SEM Philips 30 XL) incorporated in an X-ray microanalyser.

Measurements were performed at several points of the cross section of samples of medical steel AISI 316L coated by the NCD.

Figures 13, 14, 15 show the chemical constitution of elements at different points in the coating.

Concentrations of elements versus of the distance from the surface of the coating is presented in Figure 16.

On the basis of the research mentioned above, a diagrammatic model of hard carbon coatings on AISI 316L steel or titanium substrates (shown in Figure 11) has been suggested [30,31,43,45]. The surface layer several consists of carbon several thousand angstroms thick. The carbon film passes smoothly into a thick layer, a few microns) consisting of metal carbides of metals which are included in steel. This coating structure ensures a higher mechanical strength and therefore the coatings are also good protectors against corrosion in metals.

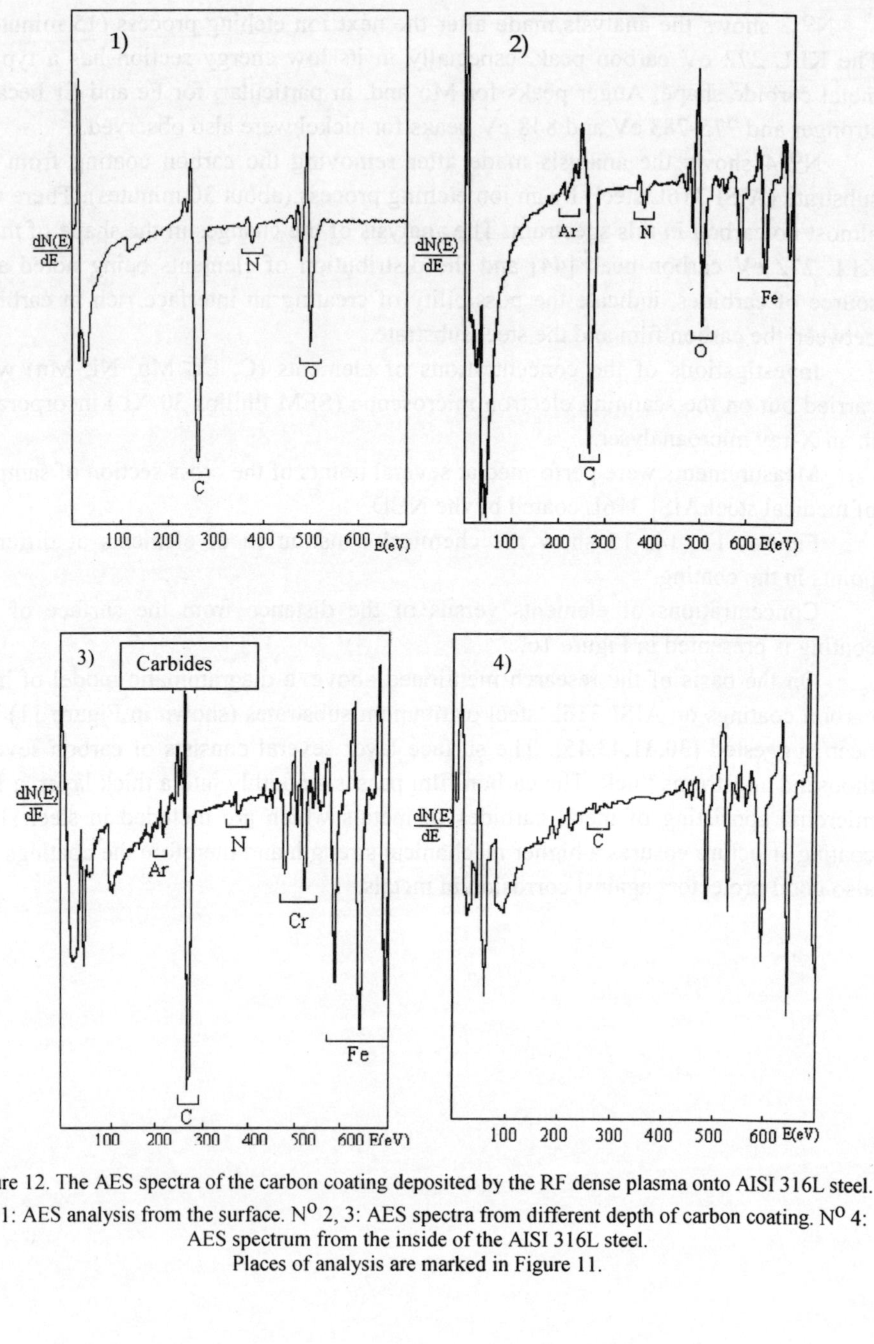

Figure 12. The AES spectra of the carbon coating deposited by the RF dense plasma onto AISI 316L steel. N^{O} 1: AES analysis from the surface. N^{O} 2, 3: AES spectra from different depth of carbon coating. N^{O} 4: AES spectrum from the inside of the AISI 316L steel.
Places of analysis are marked in Figure 11.

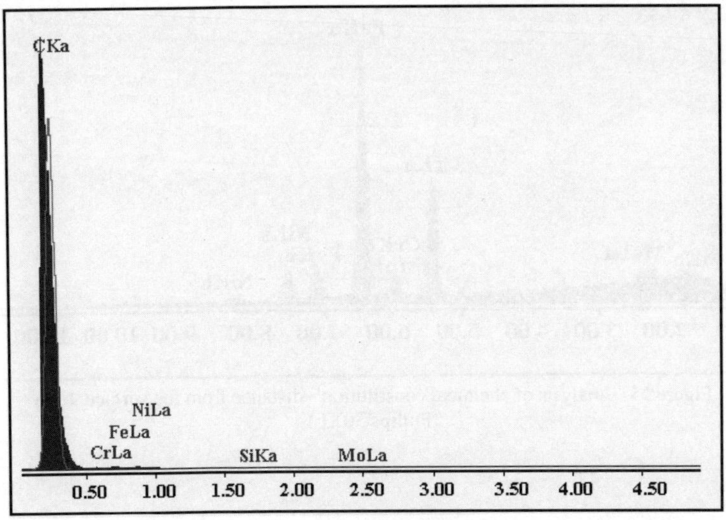

Figure 13. Analysis of chemical constitution on the surface of the coating (SEM Philips 30XL).

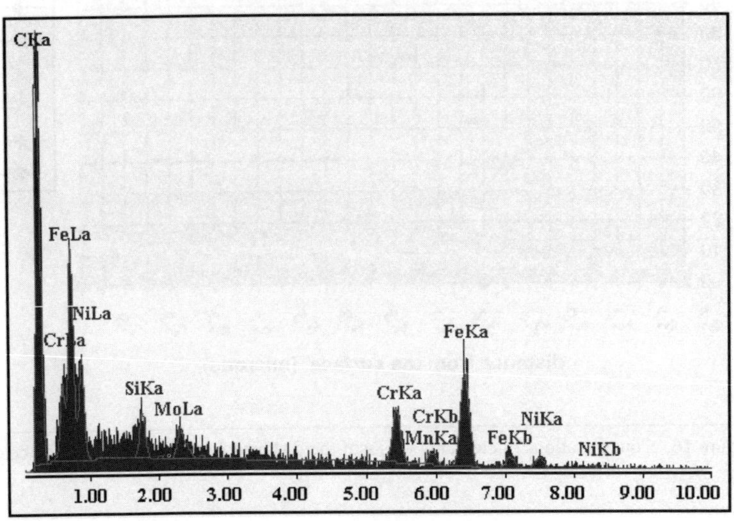

Figure 14. Analysis of chemical constitution - distance from the surface 5 μm (Philips 30XL).

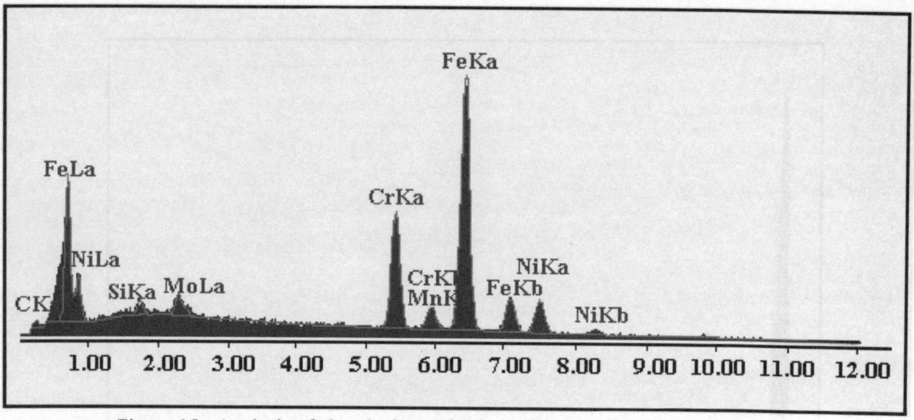

Figure 15. Analysis of chemical constitution - distance from the surface 9 μm (Philips 30XL).

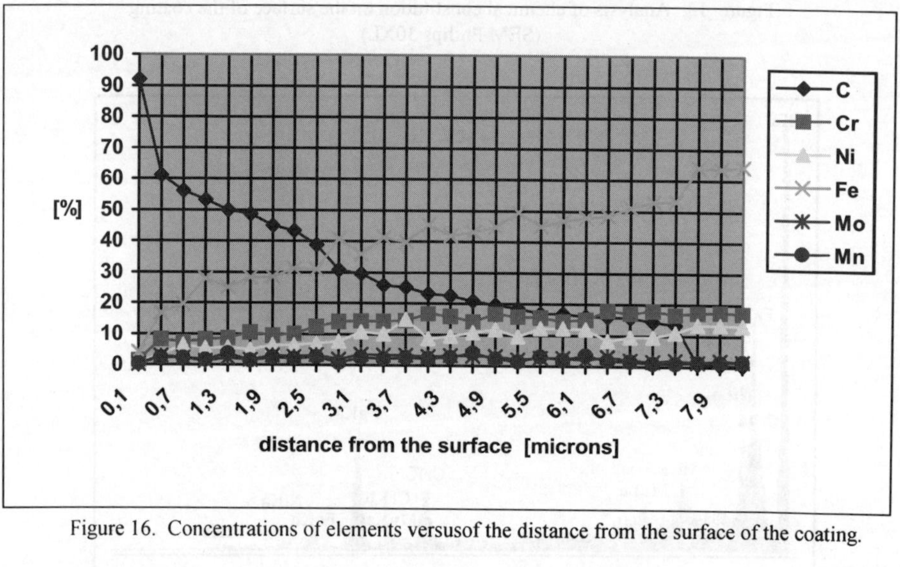

Figure 16. Concentrations of elements versusof the distance from the surface of the coating.

3.4. Tribological investigations

The results of tribological investigations are presented in Table 3. Differences between properties of NCD layers obtained by the dense plasma RF method [29] and by that applied previously by the author should be stressed. The hardness of the nanocrystalline diamond layer was equal to 95 GPa [30,31].

Table 3. Results of tribological investigations of Ti implants coated with NCD.

Parameter	*Value*
Microhardness, [GPa]	>90
Bend test (180°) (number of bending without visible changes)	2
Adhesion-scratch test (critical load), [N]	75 ± 5
Adhesion energy of the layer-substrate interface	4 Jm^{-2}
Grindability (number of screwing without visible changes)	> 100 times

The bend test through 180° (bending angle) was performed bending the thin carbon-coated titanium plate. The micrograph made by a scanning electron microscope with such a substrate after 3 subsequent bending tests shows initial cracks. The micrograph of the same substrate after 10 bending tests also shows a lot of cracking.

The grindability test was performed screwing the orthopaedic NCD-coated screws into animal bones. The examination of grindability was made by SEM after 100 subsequent screwing into the bone.

3.5. Corrosion resistance

Carbon coated specimens have been tested to find out their resistance to acids. The test consisted of one year of immersion in weak acid (HCl 2%) water solution and subsequent corrosion tests have shown that AISI 316L substrates with carbon coatings are unchanged with no corrosion products visible.

The second test is to check the chemical resistance of the coated sample. Tested samples are immersed in a chemical reagent: Mi 20 Fe (5g of copper chloride, 40 cm^3 of hydrochloric acid, 55 cm^3 of distilled water). Samples remain submerged for 15 hours in the acid before being washed in water and ultrasonic sound. Macroscopic observation shows that the surface of the uncoated steel is etched whereas the coated steel is intact. Observation performed by SEM confirms this.

These tests confirmed high chemical inertia to aggressive liquid of the AISI 316 L steel coated with carbon by the RF PA CVD method, and its potential suitability for surgical implants and other applications requiring materials with good resistance to corrosion and chemical attack.

4. Medical Applications

There are 11 elements in human organism: carbon, hydrogen, oxygen, nitrogen, sulphur, calcium, phosphorus, potassium, sodium, chlorine and magnesium. There are also 15 restigical bioelements: Fe, Zn, Cu, Mn, Ni, Co, Mo, Se, Cr, J, Fl, Sn, Si, W, As [46]. The rest of elements does not participate in metabolic reactions. The role of these bioelements is based on their activity as stimulants or on their restraining effect in activity of the enzymatic proteins which control many of the main processes of life. Too low a concentration causes inefficiency, while too high a concentration results in poisoning, sometimes even fatal (the Bertrand low [46,47]). The value of critical concentration differs for different elements and for different people.

The presence of implants in the organism and their corrosion and the resulting metalosis can be also a source of bioelements and unexpected reactions. As a consequence of corrosion, bioelements invade the tissue as metal ions or corrosion products and they can either influence the intercellular spaces or penetrate into cells (also by fagocytosis).

Toxic reactions are the most common in the scientific literature. The work of Clarke and Hickman demonstrates a correlation between corrosion decrements of different metals and alloys as well as their anodic potentials in physiological solutions and their biotolerance [46]. The lower the corrosion decrement, the higher the anodic potential which leads to better biotolerance. The biotolerance of the implant in the tissue and the fluids of the human body depends on the development of toxic and allergic processes and on mechanical reactions to the foreign body (implant). The coating process of implants by films with a high resistivity increases their biotolerance because of the bioelectronic processes. Human cells have the ability to create bonding with carbon as a main constructive element of the human body and carbon is therefore an ideal material for medical implants.

All allotropic forms of carbon meet the requirements for implants which are, for instance, biotolerance and mechanical strength. Tests of applications of different allotropic forms of carbon in medicine were made.

Carbon coatings of metallic implants or other materials to be inserted in the human body, should have the following properties.

- Biocompatibility (haemocompatibility and histocompatibility) which means the ability of human body to accept implants without destruction of the tissue.
- Chemical stability, which requires that the material is not degraded by environmental agents such as alcohol, sterilants, etc.
- Biostability, which requires that the material is not degraded by biological interactions within the body: they must not initiate reactions in tissues surrounding an implant.
- The system implant/layer must also have excellent adhesion and very good mechanical characteristics.

There have been attempts to apply all of the carbonaceous forms in medicine [6,28,30,47-58]. For example Cifre et al. [48] introduced the idea of using PCD in dentistry: the films were obtained by the HF CVD method on the ceramic substrates and used for manufacturing dental implants.

Guseva et al. [6] showed that on the bore of carbines, the new material "carbylan" can be obtained and it can be used to construct a medical implant. It is possible that carbines, with their high resistance and great strength, will replace the so-called "hard polymers" in the near future. It seems that Mitura et al. [54], Thomson et al. [56], Lettington et al [51,52]., McLaughlin et al [53], and Dowling et al. [49] have proposed the most suitable carbonaceous material for medical applications. DLC, amorphous diamond, and nanocrystalline diamond have perfect mechanical properties. The NCD has the structure formed from the hybridised bonds of the σsp^3 type, and the DLC is mixed amorphous and fine-crystalline carbon with electron hybrydisation σsp^3, $\sigma sp^2 \Pi$ and σsp^1.

The properties of the layers produced depend on several factors:

- the method of synthesis
- the parameters of the particular method
- the substrate on which the layer is deposited.

This allows for conscious choice of the deposition process and its parameters to produce carbon layers suitable for a specific application.

4.1. Pre-clinic investigation

The tests that have been undertaken so far have confirmed that the implant covered by nanocrystalline diamond is particularly suitable as a biomaterial. The mechanical properties and the results of medical tests will encourage further examination of implants with carbonaceous films. Positive results have been obtained both for steel and titanium substrates. Tests of the influence of the long- lasting environmental effect on the film i.e. that caused by live tissue, body fluid and blood, are now underway.

4.1.1. Biocorrosion resistance

Very extensive and complex investigations have been carried out for some years and these form the basis of crucial predictions for the suitability of implants made of Cr-Ni-Mo steels and coated with passive-diamond layers in clinical practice.

In general it can be said that to produce different functional forms of implants made of Cr-Ni-Mo steels, only those cast with the structure and properties which fulfil the respective ISO standards and equivalent standards should be used. Chemical composition is an important issue and an additional criterion is the level of resistance of the steel to pitting corrosion (%Cr +3.3 %Mo $\geq$ 26). A significant role in this resistance to corrosion is played by the level of steel impurities with non-metallic inclusions which is stipulated. The presence of many inclusions, especially those characterised by large size, makes it difficult to obtain a good quality of surface. During the final electrolytic polishing these inclusions are subjected to extraction, causing surface losses which facilitate the initiation of corrosion pits and lower the biotolerance.

Cr-Ni-Mo steels exhibit an austenitic structure in which $M_{23}C_6$ carbides are often observed. Their proliferation, and in particular their extraction at the boundaries of austenite grains, contribute to a decrease in the resistance to corrosion and thus heat treatment processes of steels should not be conducted in the range of temperatures at which their extraction is facilitated [46].

Implants made of Cr-Ni-Mo steels in the ground or mechanically polished state exhibit low breakdown potentials, ie approx. 400-420 mV. Implants have to be polished as the maximum roughness has been limited by their respective standards (Ra $\leq$ 0.16 μm). This roughness can be achieved through grinding, mechanical and electrolytic polishing. The traditional polishing process is performed in an electrolyte based on H_3PO_4 and CrO_3 and is followed by chemical passivation in a solution of HNO_3. Under these conditions breakdown potentials equal to 550-600 mV are obtained [46]. Marciniak and Paszenda [46,47] have developed a new electrolyte, and conditions for electrolytic polishing and chemical passivation, which facilitate the achievement of the necessary roughness by means of grinding without mechanical polishing, and ensure breakdown potentials in the range of 950-980 mV, with the current density in the passive range not exceeding 0.02 A/cm^2. A new technique for the final treatment of implants guarantees their high resistance to pitting corrosion.

Deformation of plates for fracture stabilisation with the ASIF (Association for the Study of Internal Fixation) method, i.e. through an application of the recommended deflection up to 3.5 mm, does not destroy the passive layer. Breakdown potentials remain in the range 525-725 mV [46]. Implants with the passive layer mentioned above are also resistant to stress corrosion in the Tyrod physiological solution [47].

4.1.2. Breakdown test

One of parameters which confirms suitability for medical applications is the breakdown voltage in Tyrode's solution [47]. Tyrode's solution is a water solution of salts with properties similar to those of human bodily fluids. The value of the voltage allows us to estimate the corrosion resistance inside the human body. The breakdown voltage in the Tyrode's solution for uncoated steel oscillates between 340 and 420 mV. A traditional electrolytic polishing and passivating of AISI 316L steel implants increases the breakdown voltage to 600 mV[47]. The applied coating should increase the value of the breakdown voltage further.

The deposition of nanocrystalline diamond coatings by the dense RF plasma CVD method on to medical steel (AISI 316L), results in an increase in the breakdown voltage of up to 1100−1300 mV (see Fig. 17), whereas the current density in the passive range does not exceed 0.008 mA/cm^2.

4.1.3. Investigation of biotolerance

Investigations of biotolerance were made using stainless steel AISI 316L implants, passivated and coated with carbon layer, on laboratory animals. The aim of the research was to reveal toxic and allergic reactions which occurred after placing implants in animal tissue. The research was carried out according to ASTM 981-86 standards. Guinea pigs were used as the test animals since they react to toxicological and allergic factors in a way similar to humans.

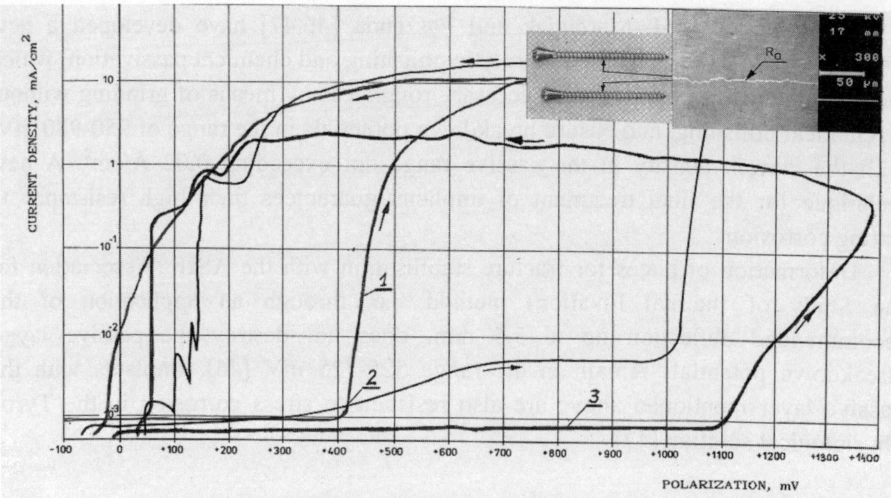

Figure 17. Curves showing the anodic polarisation of the AISI 316 steel implants in the Tyrod solution: 1 - ground surface, 2 - electrolytically polished and passivated surface, 3 - electrolytically polished and passivated surface, coated with a diamond layer.

Implants in the form of disks, 5.0 mm in diameter and 1.0 mm thick, of mass equal to 0.750 ± 0.010g, were investigated. Disks were polished electrolytically and fitted and were coated with a carbon layer. Disks were implanted into chest muscles, subcutaneously between shoulder blades and into the tibia bone. Investigations were carried out for 24 to 52 weeks.

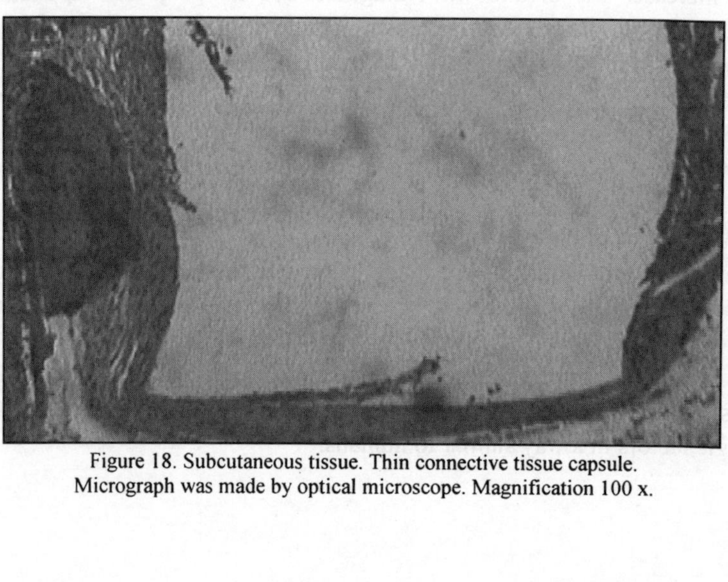

Figure 18. Subcutaneous tissue. Thin connective tissue capsule. Micrograph was made by optical microscope. Magnification 100 x.

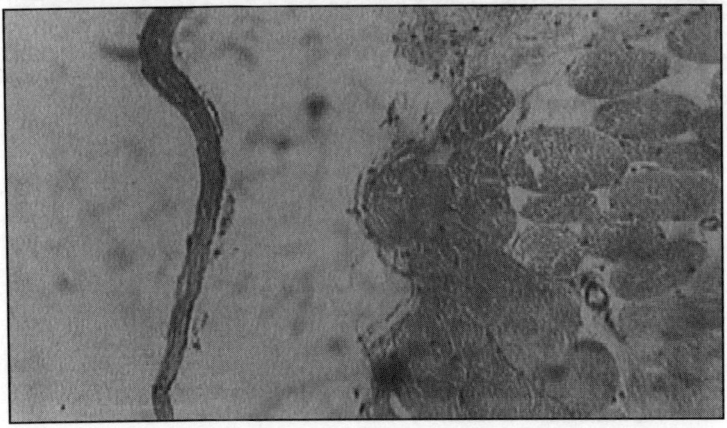

Figure 19. Muscles. Thin connective tissue capsule adhering to loose connective tissue. The micrograph was made by optical microscope. Magnification 100 x.

After the implantation of samples the animals quickly returned to normal and there were no noticeable differences in their behaviour. X-ray investigations showed that implants had been inserted correctly. No reaction was seen, nor were any changes found around the implants in the x-rays.

On the basis of this research it was found that in subcutaneous tissue, muscles and bones, thin connective tissue capsules built from fibrocytes and collagen fibres were formed. In the wall of the capsule neither a phagocytic reaction nor products of corrosion were seen.

Internal organs (liver, kidneys, spleen) did not show any pathomorphological changes. Observations were made by optical microscope (see Figures 20 and 21).

Histopathological investigations showed a very good biotolerance of implants coated with the NCD layers. The coating protects an implant efficiently against corrosion and metalotoxemie.

4.2. Examples in medicine

Results of preliminary histopathological investigations on laboratory animals were so promising (there were no changes in any of the animals after several months) that the prognosis for implanting prostheses in humans is good.

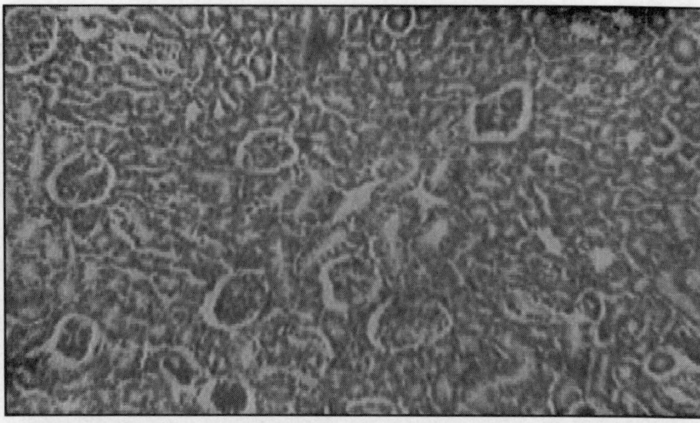

Figure 20. Kidney, without any pathomorphological changes after 52 weeks. The micrograph was made by optical microscope. Magnification 100 x.

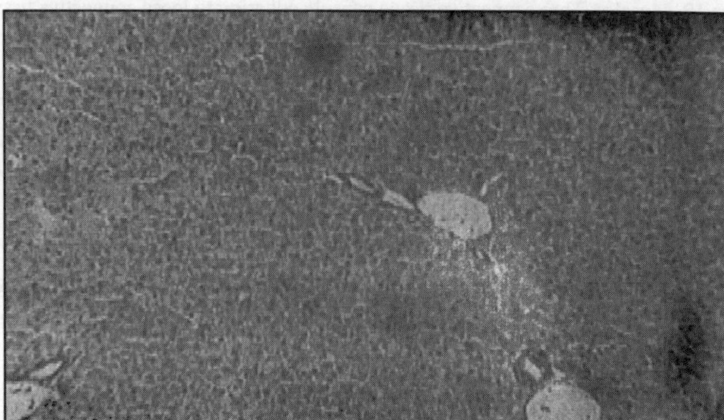

Figure 21. Liver, without any pathomorphological changes after 52 weeks. The micrograph was made by optical microscope. Magnification 100 x.

So far, carbon-coated implants have been used at many medical centres in Poland and abroad, such as those in Gliwice, Łódź (Poland) and Liberec (Czech Republic).

The paper in the Journal of Chemical Vapor Deposition [58] presents a sample case of *Pseudoarthrosis of the left femur. Ostisis following treatment of an open fracture,* treated at the Orthopedic-Trauma Ward of the Biernacki Hospital in Pabianice.

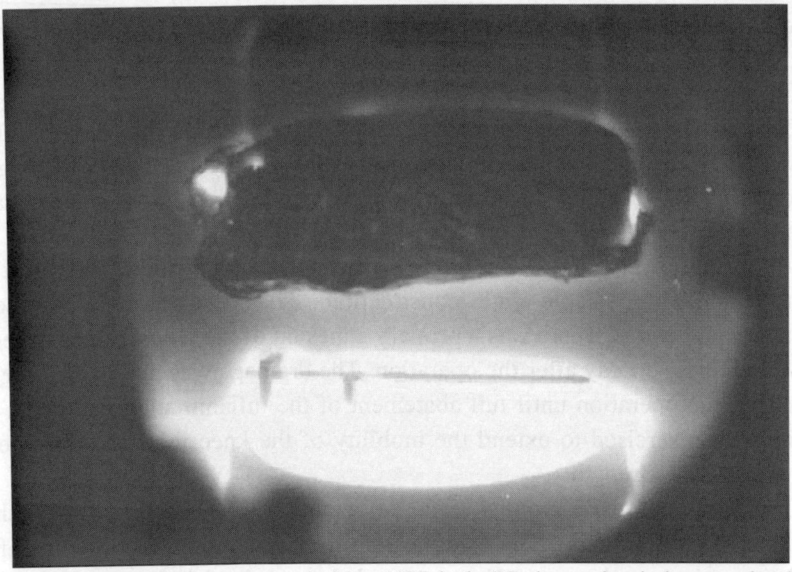

Figure 22. The Schantz implant coated with an NCD in the RF plasma-chemical reactor chamber.

This example shows a case of *Pseudoarthrosis of the left femur. Ostisis following treatment of an open fracture,* treated at the Orthopedic-Trauma Ward of the Biernacki Hospital in Pabianice near Łódź (Poland). The patient, 50 years old, was admitted in April 1992 with an open multiple fracture of the left femur, a closed fracture of the shaft of the right femur, and an open double fracture of the left tibia. Both bones were operated on on the day of admission with the use of Kuntscher rods. The post-operative course was complicated because of festering wounds in the left femur. The installing of drainage from the soft tissues resulted in the healing of the wound by granulation. A few days later, a pus-dripping fistula appeared below the scar. The patient was operated on again and the Kuntscher rod was removed, the through periosteous drainage was installed, and the limb was immobilized. The local condition did not improve despite the operation. The patient was given preventive treatment: a daily change of dressings and rinsing of the wound with solutions of antibiotics.

The intensification of the inflammation process, with the fistula secreting a large amount of pus, necessitated another operation. A sequestrectomy was performed, with the installation of an intraosseous drainage and the resection of the fistula and necrotic tissues. The operating team installed a stabilizing Martin apparatus and introduced two Schantz implants, coated with an NCD (compare

Figure 22), centrally threaded in the frontal plane, and further two Schantz implants coated with an NCD, unilaterally threaded in the fibular plane into the distal fraction.

Two NCD-coated Schantz implants unilaterally threaded into the frontal plane and two NCD-coated Steiman implants in the plane deflected by 45 degrees from the frontal plane were introduced into the proximal fraction in the sub-trochanteric region.

All of these were joined externally by means of carbon rods. The result was a biplanar stabilizing "staple-tent" construction. The post-operative course was without complications. There was a primary adhesion of the wound and sutures were removed on the 14th day after the operation. The drainage was maintained for seven weeks after the operation until full abatement of the inflammation had taken place. The limb was exercised to extend the mobility of the knee joint and the talocrural joint, and muscle strength.

The X-ray control 3 months after the operation (compare Figure 23) showed the fusion of the fractions. The external stabilizer was removed 9 months after its installation, when a subsequent X-ray control showed the full fusion of the infected false joint and the new bone. The 40-degree flexion of the knee joint is caused by the scarring of soft tissues resulting from the prolonged inflammation and an 18-month period of immobilisation.

5. Mechanical Applications

The examination described in this chapter was made in Laboratorie de Materiaux in ECAM (L`Ecole Catholique d`Arts et Mètiers), Lyon. It was preceded by an investigation involving covering of cemented carbide cutting tools with wear-resistant carbon films examined in the Technical University of Lodz and in the Warsaw University of Technology. The work concerned three types of cutting tool: slitting saws, gang-type cutters and cutter head plates. In each case the life of the tool was extended by 50 to 200 percent, depending on the types of tool and parameters of deposition. All milling tests were carried out in real work conditions in the wood industry. Detailed information about the properties of a carbon film, an NCD (nanocrystalline diamond) and DLC (diamond-like carbon) sintered in the RF PCVD method, and wear behaviour of cutting tools with and without a carbon layer, was described in works by Niedzielski, Sokolowska, Miklaszewski, Grabarczyk, Golabczak and others [59-61].

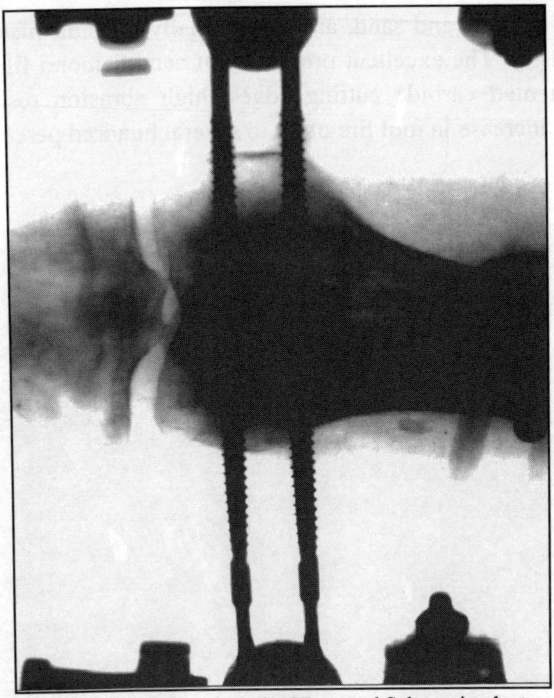

Figure 23. X-Ray image of NCD coated Schantz implants inside of the bone after few months of implantation.

Cutting tools have an important role to play in modern machines and appliances, as well as in utilitarian objects. Constantly increasing requirements for high quality surfaces require ever more sophisticated research. One of the possible directions of research into increasing wear resistance and the durability of tools consists of coating the surface with thin films of either metals or metal compounds, such as TiC, WC, TiN. Thin and hard layers are being deposited by means of PVD and CVD methods.

An attractive method of enhancing the durability of cutting tools is to coat their working surface with carbon and diamond-like films [60,61]. With the RF PCVD method (radio frequency chemical vapour plasma deposition), thin NCD films are deposited on cemented carbide tools (see Fig. 24) used for cutting particle boards made from wood and wood-based materials. In comparison with tools used for machining metals, tools for machining wood and wood-based materials are subjected to different types of wear mechanism arising from the higher processing speeds and the lack of cooling agents. In addition wood-based materials, compounds of wood, melamine resin and sand, are exceptionally difficult materials for milling in the wood industry. The excellent properties of nanodiamond films, such as good adhesion to cemented carbide cutting edges, high abrasion resistance and high hardness, give an increase in tool life of up to several hundred percent.

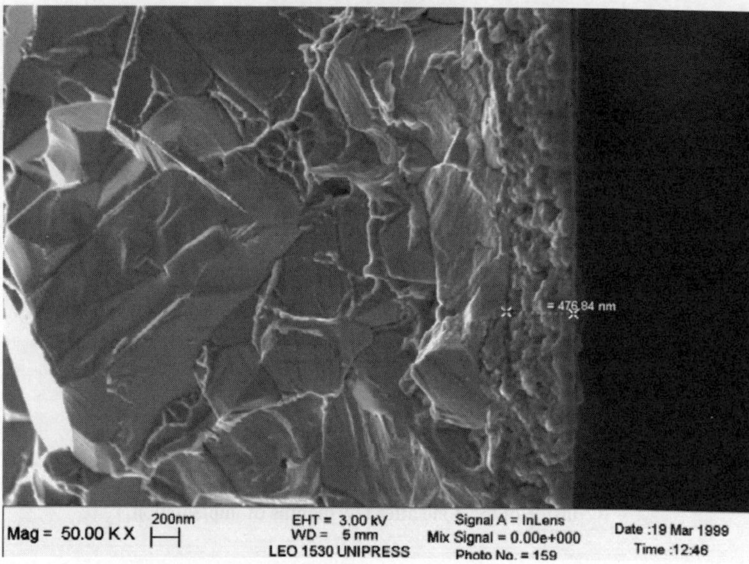

Figure 24. NCD coated cutting tool from cemented carbide.

Studies of diamond-like films deposited onto cemented carbide tools gave the answers needed to work out optimal methods of synthesis of these wear resistant coatings on edges of cutting tools for wood-based materials. A new method of locating substrates in the working chamber during the process by using a special type of holder, permitted researchers to achieve equal films on all lengths of cutting edge, with good mechanical properties.

The characteristics of wear of the metal slitting saws investigated have been presented in Fig.25. The lifetimes of metal slitting saws, determined for the assumed values of permissible wear VB_{dop} =0.25/, are: T_1=58 min - for a metal slitting saw subjected to ordinary thermal treatment; and T_2=520 min - for a metal slitting saw covered with a carbon coating. The results obtained confirm almost a ninefold increase in the life of a metal slitting saw with a carbon coating. A hard coating of carbon on the working surface of the saws improves their functional quality to a remarkable extent.

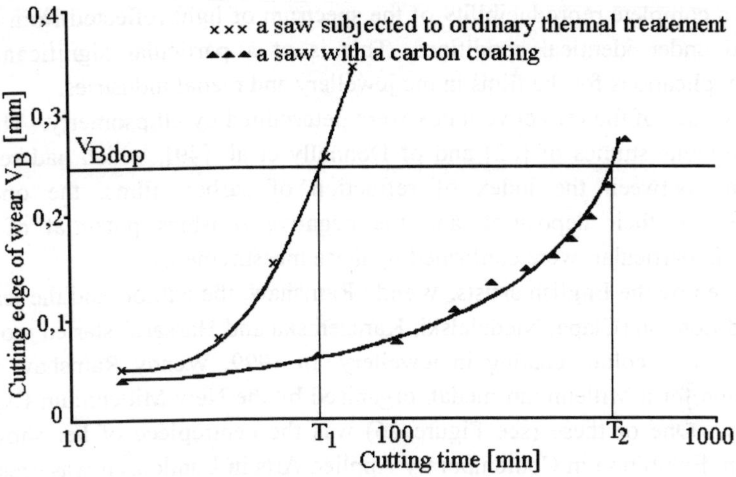

Figure 25. The course of wear of metal slitting saws.

Great demand for this type of tool, stemming from the dynamic development of the furniture industry, quality improvement in its products and a need for ways to cut production costs, is an encouraging factor that makes the author continue this research and plans to work on further improvement of the technology of the NCD film deposition on to cutting tools for machining woody materials.

Creating a coating consisting of a mixture of carbon, iron and titanium gave interesting results. With this method it should by possible in the future to sinter carbon films with an admixture of other elements to change mechanical and physical properties. The next step will be to find new applications for carbons and nanodiamond films deposited in the RF PCVD method.

6. Applications in Jewellery

Thin carbon films for application in jewellery manufacturing were synthesised by the decomposition of a hydrocarbon in a radio frequency plasma [30,54], at a pressure ranging a few to several dozen Pascals, with an accompanying negative self-bias potential of several hundred volts. The films manufactured in this way were studied in order to determine their structure, their optical properties and their mechanical and chemical strength.

Colourimetric analysis, consisting of the assessment of the colour quality, revealed a complete reproducibility of the spectrum of light reflected from the films deposited under identical conditions. This is of a particular significance when seeking applications for the films in the jewellery and medal industries.

The values of the refractive index were determined by ellipsometry. The results of the previous studies of [62] and of Donnelly et al. [49], which had revealed a correlation between the index of refraction of carbon films, the operational parameters of their deposition and the negative self-bias potential of the hot electrode in particular, were confirmed by these measurements.

Inspired by the English artists, Wendy Ramshaw, the author, and the team from Lodz and London (Clapa, Niedzielski, Karczemska and Hassard) started work in the field of carbon colour coating in jewellery. In 1999, Wendy Ramshaw won the competition for a Millennium medal, organised by the New Millennium Experience Company. One of these (see Figure 26) was the centrepiece of her show at the Millenium Exhibition in Contemporary Applied Arts in London. It was presented to Her Majesty the Queen on 31 December 1999 for the celebrations at the opening of the dome.

Other medals were designed for the Prime Minister, and Heads of State who visited the Millennium Dome at Greenwich (Figure 27).

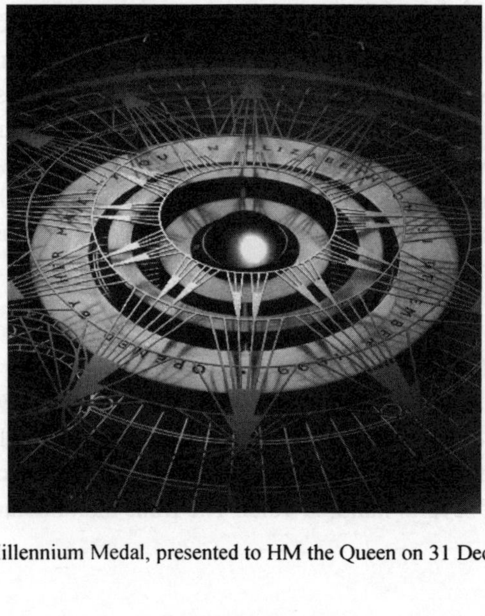

Figure 26. Millennium Medal, presented to HM the Queen on 31 December 1999.

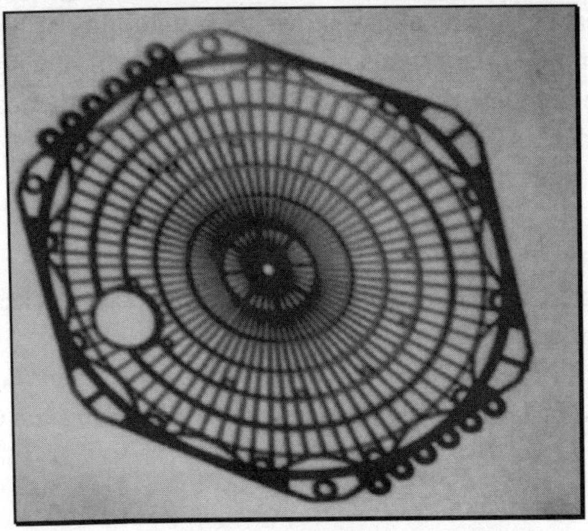

Figure 27. Millennium Medal, coated by DLC with blue colour.

7. Conclusions

The coating deposited on AISI 316L steel or on titanium consisting of ultrafine-grained diamond bound to the substrate by carbides seems to be an excellent biomaterial.

The present stage of investigation allows us to draw optimistic conclusions concerning the potential application of carbon coatings obtained by the dense RF methane plasma method.

Naturally the specific nature of the investigations and medical tests requires long-lasting observation before the suitability of coatings can be determined, but this material already looks very favourable.

The use of carbon-coated implants protects them from biological corrosion, metalosis, and prevents the creation of local cells. This has a stabilizing effect and makes the recovery of a patient possible even in serious cases.

In same cases, the use of metal implants without carbon coatings makes the recovery impossible. In each case, the use of the carbon coatings will shorten the treatment period.

Films of nanocrystalline diamond, deposited on cutting tools made of cemented carbides for machining wood-like materials, bring about an improvement in the mechanical properties of these tools as well as giving rise to a significant increase in their durability.

Colour carbon coatings have been developed for a range of substrates and the potential for replicating such coatings shows their suitability as a material for use in jewellery manufacture. They can also be used in a wide of other applications where colours need to be combined in a tough, hard-wearing, aesthetically appealing and biocompatible product.

Acknowledgements
The author is indebted to include many thanks for Professor Aleksandra Sokolowska for enthusiastic support and friendship.
Works were financed by the Polish State Committee for Scientific Research (KBN) and were supported by projects of the British Council and French Government.

References

1. Goresy A.El, Donnay G.T., *Science* **161** (1968), 363-366.
2. Whittaker A.G., *Science*, **200 N 4343** (1978), 763-764.
3. Gilkes K.W.R. and Pillinger C.T., in *Carbyne and Carbynoid Structures,* eds. Heimann R.B., Evsyukov S.E. and Kavan L. (Kluwer Academic Publishers, Dordrecht, Boston, London 1999), 17-30.
4. Kasatoczkin V.I. et al., *DAN SSSR* **214** (1974), 887-890; *Carbon* **11** (1973), 70.
5. Kudryavtsev Yu.P., in *Carbyne and Carbynoid Structures,* eds. Heimann R.B., Evsyukov S.E. and Kavan L. (Kluwer Academic Publishers, Dordrecht, Boston, London 1999), 1-7.
6. Guseva M.B., Babaev V.G., Kudryavtsev Yu.P., Alexandrov A.F., Khvostov V.V., *Diamond Related. Mater.* **4** (1995), 1142-1144.
7. Sokolowska A. and Olszyna A., in *Carbyne and Carbynoid Structures,* eds. Heimann R.B., Evsyukov S.E. and Kavan L. (Kluwer Academic Publishers, Dordrecht, Boston, London 1999), 117-132.
8. Kratschmer W., Lamb L.D., Fostiropoulos K. And Huffman D.R., *Nature* **347** (1990), 354-358.
9. Kroto H.W., Heath J.R., O'Brian S.C., Curl R.F. and Smalley R.E., *Nature* **318** (1985), 162-166.
10. Curl R., Smalley R., *Pour la Science*, **170, XII**, (1991).
11. Iijma S., *Nature* **354** (1991), 56-57.
12. Bundy F.P., Hall H.T., Strong H.M., Wentorf R.J., *Nature* **176** (1955), 51-54.

13. Angus J., Koidl P., Domitz S., in *Plasma Deposited Thin Films,* eds. Mort J., Jansen F. CRC Press, Inc., Boca Raton 1986.
14. Spitsyn B., Derjaguin B., Patent SSSR No. 339134, (1980).
15. Bachmann P.K., Messier R., *Special Report C&EN* **May 15** (1989), 24-39.
16. Kamo M., Sato Y., Matsumoto S. and Setaka N., *J. Crystal Growth* **62** (1983), 642-644.
17. Mitura S., *J. de Physique IV (*suppl. to *J. de Physique II)* **1, C2** (1991), C939-C944.
18. Mitura S., *Diamond Related Materials* **1** (1992), 239-242.
19. Boudina A., Fitzer E., Wahl G., *Diamond Rel. Mater.* **1** (1992), 380-387.
20. Bubenzer A., Dischler B. and Nyaiesh A., *Thin Solid Films* **91** (1982), 81-87.
21. Dischler B., Bubenzer A. and Koidl P., *Solid State Comm,* **48,** (1983), 105; *Appl. Phys. Lett.* **42** (1983), 636-638.
22. Mitura S., *The role of electrons in low-pressure diamond synthesis* (Scientific Bulletin Technical University of Lodz, No. 666, Lodz 1992).
23. Enke K., Dimigen H. and Hubsch H., *Appl. Phys. Lett.* **38** (1980), 291-292.
24. Vora H., Moravec T., *J. Appl. Physics* **52,** (1981), 6151.
25. Holland L., Patent U.K. No. 1582231, (1976).
26. Holland L., Ojha S.M., *Thin Solid Films* **48** (1978), L21-L23.
27. Has Z., Mitura S., Wendler B., in *Proc. International Ion Engineering Congress: Int. Symposium on Ion Sources and Ion Assisted Technology and Internat. Conf. on Ion and Plasma Assisted Techniques,* ed. Takagi T. (Kyoto, 1983), 1143-1144.
28. Mitura E., Mitura S., Niedzielski P., et al., *Diamond Rel. Mater.* **3** (1994), 896-898.
29. Mitura S., Mitura E., Mitura A., *Diamond Rel. Mater.* **4** (1995), 302-303.
30. Mitura S., et al., in *Proc. 8th CIMTEC World Ceramic Congress & Forum on New Materials; "New Diamond and Diamond-Like Films", ADVANCES IN SCIENCE AND TECHNOLOGY, vol. 6;* ed. Vinzenzini P. (TECHNA srl., Faenza 1995), 273-280.
31. Mitura S., in *Proc. SPIE, Solid State Crystals in Optoelectronics and Semiconductor Technology,* eds. Rogalski A., Rutkowski J.,. Majchrowski A., Zielinski J., **3179** (1997), 79-86.
32. Thompson L.F., Mayhan K.G., *J. Appl. Polymer Sci,* **16** (1972), 2317-2326.
33. Grebowicz J., Pakula T., Wrobel A.M., Kryszewski M., *Thin Solid Films* **65** (1980), 351-359.
34. Mitura S., *J.Crystal Growth* **80** (1987), 417-423.

35. Coburn J.W., Kay E., *J. Appl. Physics* **43** (1972), 4965-4969.
36. Davidse P.D., *Vacuum* **17(3)** (1966), 139-145.
37. Mitura S. et al., *High Technology vol. 38: Diamond Based Composites and Related Materials,* eds. Prelas M.A. et al. (NATO ASI Ser., Kluwer Academic Publishers, Dordrecht, London 1997), 211-228.
38. Dusek V., Vanecek M., Siroky P., Vorlicek V., *Diamond Related Mater.* **2** (1993), 397-401.
39. Hiller T., RENISHAW - Raman Spectra, DIAMOND'96, Tours, France 1996.
40. Mermoux M., Roy F., Marcus B., Abello L., Lucazeau G. *Diamond Rel. Mater.* **1** (1996), 798-801.
41. Oral B., Hauert R., Muller U., Ernst K.-H., *Diamond Rel. Mater.* **4** (1995), 482-487.
42. Stuart S.-A., Prawer S., Weiser P.S., *Diamond Rel. Mater.* **2** (1993), 753-757.
43. Rylski A., Niedzielski P., Mitura S., *Journal of Chemical Vapor Deposition,* **3**(4) (1995), 296-301.
44. Chang Chuan C., in *Characterization of Solid Surfaces,* eds. Kane P.F. and Larrabee G.R. (Plenum Press, New York, 1974), 535.
45. Olborska A., Swider M., Wolowiec R., Niedzielski P., Rylski A., Mitura S., *Diamond Rel. Materials* **3** (1994), 899-891.
46. Marciniak J., *Biomaterialy w chirurgii kostnej* (Wyd. Politechnika Slaska. Gliwice, 1992).
47. Marciniak J., Boba J., Paszenda Z., Mitura S., *Werkstoffe und Korosion* **44** (1993), 379-383.
48. Cifre J., Polo M.C., Sánchez G., Lousa A., Esteve J., *Diamond Related. Mater.* **4** (1995),798-801.
49. Donnelly K., Dowling D.P., McConnell M.L. and Flood R.V., *Diamond and Related Mater.* **8** (1999), 538-540.
50. Gril A., Patel V., *Diamond Rel. Mater.* **2** (1994), 597-605.
51. Lettington A.H., *Philos. Trans. R. Soc. London* **Ser. A, 342** (1993), 287-296.
52. Lettington A.H., Smith C., *Diamond Related. Mater.* **1** (1992), 805-809.
53. McLaughlin J.A., Meenan B., Maguire P., Jamieson N., *Diamond Related. Mater,* **5** (1996), 486-491.
54. Mitura S., Has Z. and Gorokhovsky V.I., *Surface Coatings Technology* **47** (1991), 106-112.
55. Sleptsov VV., et al., *Diamond Related. Mater.* **5** (1996), 483-485.
56. Thomson L.A., Law F., Rushton N., Frank J., *Biomaterials* **12** (1991), 37-40.

57. Tochitsky E.I., Selifanov O.V., Stanishevsky A.V., Akulich V.V. and Kapustin I.A., *Surface and Coat.Technol.* **47** (1991), 792-796.
58. Zolynski K., Witkowski P., Kaluzny A., Has Z., Niedzielski P., Mitura S., *Journal of Chemical Vapor Deposition,* **4**(1) (1996), 253-258.
59. Gołabczak A., Niedzielski P., *Journal of Chemical Vapor Deposition* **3/4** (1996), 332-341.
60. Niedzielski P., Sokolowska A. et al., *Diamond Rel. Materials* **10** (2001), 1-11.
61. Morita T., Banshoya K., Tsutsumoto T., Murase Y., *Mokuzai Gakkaishi* **41** (12) (1995), 1093-1101.
62. Mitura S., Sokołowska A., Szmidt J., Balasiński A., in *Applications of Diamond Films and Related Materials,* eds. Tzeng Y., Yoshikawa M., Murakawa M., Feldman A. (Elsevier Sci. Publ. B.V., Amsterdam, 1991), 795.

CHAPTER FIVE

CCVD: LOW-COST VAPOUR DEPOSITION OF THIN FILMS IN THE OPEN ATMOSPHERE

Andrew T. Hunt

MicroCoating Technologies, Inc., 3901 Green Industrial Way
Chamblee, GA 30341, USA; Email: ahunt@microcoating.com

1. Introduction to the Combustion CVD Process

The innovative Combustion Chemical Vapour Deposition (CCVD) process, patented[1] for applications other than diamond coatings by the Georgia Institute of Technology and licensed exclusively to MicroCoating Technologies, Inc., (MCT, Chamblee, GA) has demonstrated its ability to overcome many of the shortcomings of traditional vapour deposition techniques while yielding equal or better quality coatings at a lower cost. The key advantage of the CCVD technology is the ability to deposit thin films in the open atmosphere (or any pressure greater than 10 kPa) from inexpensive precursors. Furthermore, there is no need for costly furnaces, vacuum equipment, reaction chambers, and post-deposition treatment (e.g. annealing).

A schematic representation of a CCVD system and its major components is given in Fig. 1. Typically, precursors are dissolved in a combustible solvent which also acts as the fuel for the flame. This solution is atomized to form submicron droplets by means of a proprietary technology (Nanomiser™ technology, patent pending). These droplets are then carried by an oxidizing gas stream into a flame where they are combusted. Coating is accomplished by simply drawing combustion plasma containing the activated deposition species over the substrate's surface. The thermal energy from the flame provides the means to evaporate the droplets and for the precursors to react and deposit on to the substrate. The entire process generally takes no more than two hours from initial set-up to post-deposition cleaning. Multiple depositions utilizing the same solution require less than 30 additional minutes per run. Therefore, coatings and their properties can be optimized quickly in a systematic manner.

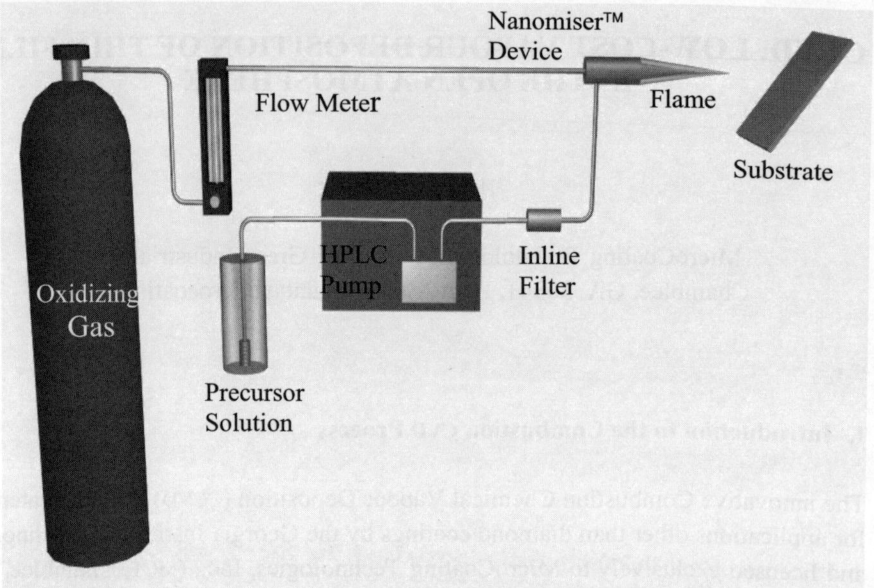

Figure 1. Schematic representation of the open-atmosphere CCVD system showing its major components.

The liquid solution CCVD process does not deposit droplets (these evaporate in the flame environment) or powders as in traditional thermal spray processes. The CCVD technology is different from spray pyrolysis in which a liquid mixture is sprayed on to a heated substrate, while CCVD atomizes a precursor solution into sub-micron droplets followed by vaporization of those droplets. The capabilities and properties of the resulting coating demonstrate that CCVD is a true vapour deposition process. For example, depositions are not line-of-sight limited and achieve epitaxy; 10 nm thick dielectric coatings onto silicon wafers in a Class 100 clean room resulted in a capacitor yield of 100%. This result attests to the high quality of the deposited coatings which were found to be free of particles, pinholes and other defects. For some materials, substrate temperatures may be as low as 100 °C, thus allowing deposition onto a wide variety of materials including polymers.

By adjusting solution concentrations and constituents, a wide range of coating stoichiometries and compositions can be achieved. This is especially valuable for achieving specific compositions and thin film characteristics. Conventional CVD requires precursors with sufficiently high vapour pressures to enable vapour phase transport. This frequently necessitates the use of expensive materials and often produces toxic fumes which must be carefully treated ("scrubbed"). In contrast, the CCVD process uses soluble precursors that do not need to have a high vapour pressure. Precursors for the CCVD process tend to be between 10 and 100 times less expensive than those used in traditional CVD processes. The physical structure and chemical composition of the deposited films can be tailored to the specific application requirements which greatly facilitates the rational design of thin films.

In summary, the CCVD process offers the following capabilities:

- **No need for a specialized chamber.** Depositions are inexpensively performed in ambient conditions without the need for expensive, specialized equipment such as reaction furnaces and/or vacuum chambers, reducing repair times and costs.
- **Uses inexpensive precursors.** Soluble precursors are used instead of expensive, high vapour pressure organometallics.
- **Excellent composition control.** Solution composition is adjustable to allow great versatility in depositing a wide variety of complex multi-component compounds with targeted stoichiometries.
- **Suitable for continuous production.** The equipment facilitates a robust production system which has been shown to operate without failure working round the clock for several days. Substrates that have been coated include fibre tows, pins, industrial rollers, wire, radiators, and roll sheet material.
- **Straightforward integration into existing production processes.** Advanced materials can be deposited onto large surface areas, including assembled parts that would be difficult to place inside a traditional coating chamber.
- **Ability to deposit multi-layered structures.** CCVD flames are regularly used in sequence to deposit multi-layered structures.
- **Wide choice of substrates.** The CCVD process allows deposition onto many different substrate materials, including plastics which can be incompatible with vacuum processes. Different sizes and shapes can be processed including non-planar substrates.
- **Precise control of coverage area.** Coating can be limited to specific areas of a substrate by simply controlling the dwell time.
- **No line-of-sight limit.** Operating at atmospheric pressure randomizes atom trajectories through diffusion. This permits infiltration into high

aspect ratio holes and vias, as well as complete coverage around fibres and wires. This contrasts with conventional long mean-free-path PVD processes where the "shadowing" of molecular rays affects distribution of coatings over non-planar surfaces.

- **Outstanding microstructure control.** - Microstructure of the deposited film is closely controlled, ranging from dense, *epitaxial* films to high surface area, nanoporous layers.
- **Accelerated development cycle for new applications.** Development of coatings for specific applications is achieved more rapidly than with traditional technologies: a large number of samples are quickly prepared for testing and optimization.
- **Environmentally friendly.** Relatively safe chemical precursors (non-toxic and halogen-free) are used, resulting in benign by-products, thus reducing environmental impact.

As a result, capital requirements and operating costs are reduced up to tenfold when compared with competing chamber- and vacuum-based technologies (e.g. sputtering and most CVD). The ability to deposit thin films in the open atmosphere enables continuous, production-line manufacturing. Consequently, throughput potential is far greater than with conventional thin-film technologies, most of which are generally restricted to batch processing. So far CCVD has been used to prepare more than seventy distinct material compositions for a variety of applications.

2. Case Studies

This section highlights some of the process capabilities of CCVD.

2.1. Capacitor coatings[2]

There is worldwide interest in developing processes capable of manufacturing high permittivity (high k) thin films for IC applications. The CCVD process is very suitable for high-rate deposition (up to 1 μm/min) of dielectrics such as SiO_2. Multi-component perovskite thin films (e.g. barium strontium titanate, BST) with extraordinarily high dielectric constants have also been successfully deposited.

Thin films comprising BST are very difficult to prepare by any competing thin film process. These films not only hold promise for small-area embedded capacitors, but also have significant potential for on-chip applications.

Additional work on dielectrics includes deposition of silica, lead aluminum boron silicates (PABS), lead lanthanum zirconium titanate (PLZT) and BST on Si-Ti-Pt wafers (Fig. 2). The wafer specimens were patterned with metal electrodes and electronic properties were characterized. Thickness, and correspondingly capacitance variation, was less than 2%. The absence of impurity peaks in XPS spectra of silica-coated specimens demonstrates the achieved purity. Yield, defined as the percentage of functioning vs. total measured capacitors, was 100%.

Breakdown field strength was in the range 1.1-5.4 MV/cm and leakage current density was about 10^{-7}-10^{-9} A/cm^2 at 0.5 MV/cm. Capacitance density was 23-350 nF/cm^2, depending on thin-film thickness and materials. No breakdown was observed after 20 cycles between 0-40 V. Time-dependent dielectric breakdown (TDDB) was 185 s at 40 V for ten of the patterned capacitors.

For capacitors on copper foil substrates, yield of initial samples was an impressive 83-90%. Capacitance density ranged from 20-242 nF/cm^2 which is approximately 50-200 times higher than that of laminated planar epoxy-glass capacitors.

2.2. *Coating conformality*[3]

Many applications require the deposition of *conformal* coatings, however, due to the micro-roughness of many substrates, line-of-sight deposition processes (e.g. PVD processes) can suffer from "shadowing" of peaks. Shadowing, which results from long mean-free molecular paths, can cause pinhole defects in the resulting films. In contrast, short molecular mean-free-paths, characteristic of the CVD process, effectively randomize the path trajectories of the atoms, thus minimizing shadowing. The coating effectively conforms to the topography of the substrate surface. An example is the conformal coating of over a thousand 12 μm diameter alumina fibres (NEXTEL™ 610) with $LaPO_4$ in a deposition at 900 °C. More than 50% of the fibres exhibit a coating thickness from 300-500 nm, with no uncoated fibres.

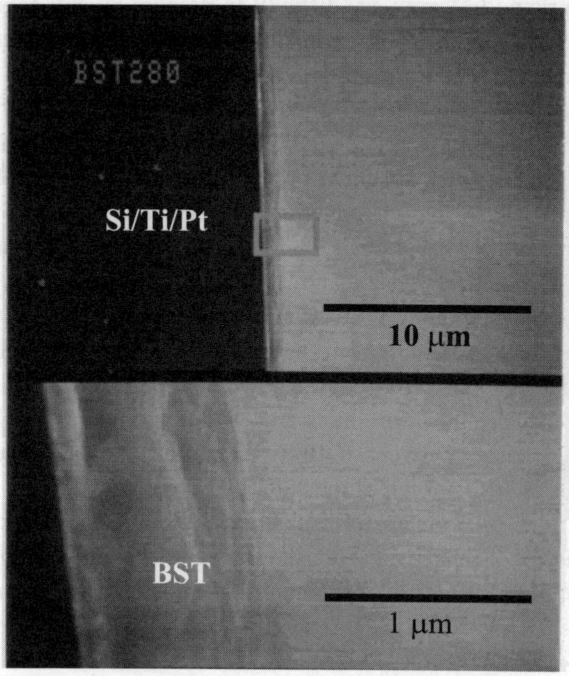

Figure 2. SEM micrograph of dense thin film of BST on an Si/Ti/Pt substrate. The lower picture is a 10× magnification of the framed area in the upper picture.

2.3. Epitaxial coatings[4]

Epitaxial $SrTiO_3$ was successfully deposited onto a MgO single crystal (Fig. 3). A single cube-on-cube texture was displayed in pole figure X-ray diffractometry. The full width at half-maximum (FWHM) values of the (110) and (200) planes are 2.020° ± 0.004° and 1.67° ± 0.01°, respectively. According to Oak Ridge National Laboratory (ORNL), these values are the best ever reported for a $SrTiO_3$ coating deposited on this substrate by any technique. The approximately 380nm thick epitaxial coating is smooth and dense.

Cerium oxide (CeO_2) was deposited onto a (100) $LaAlO_3$ single-crystal substrate with a single cube texture and good in- and out-of-plane alignment. A φ scan of the (220) plane of the CeO_2 film yielded a FWHM of 2.45° ± 0.05°; the rocking curve of the (200) plane indicated a FWHM of 1.688° ± 0.006°. A layer of

YSZ was then deposited by CCVD on top of the ceria layer. The majority of the YSZ deposited as one cube texture with only a minor 45° film rotation (<1% intensity) present. The YSZ film exhibited good in- and out-of-plane alignment with a FWHM

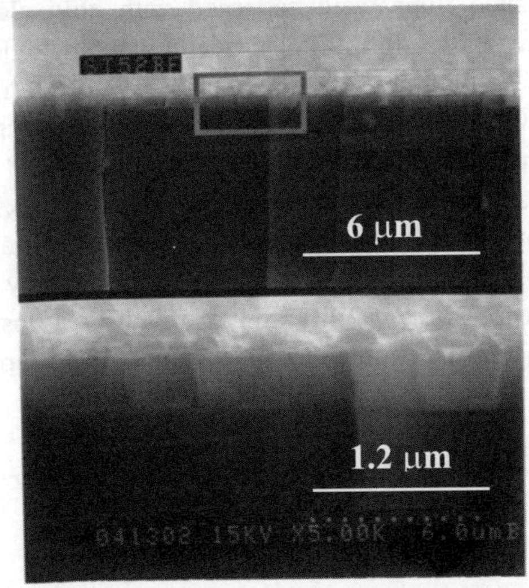

Figure 3. An epitaxial layer of strontium titanate was successfully deposited onto a MgO single crystal. The lower picture is a 5× magnification of the inset in the upper picture.

from the (220) plane of 3.37° ± 0.01° and a FWHM from the (200) plane of 2.37° ± 0.01°. Numerous other materials have also been deposited by CCVD as epitaxial films, for example Ta_2O_5, PLZT, BST, YBCO, LSC, and YSZ.

2.4. Frequency agile thin films for wireless communications[5–7]

Recent advances in semiconductor IC fabrication technologies have stimulated research and development of integrated ferroelectric oxide thin films which can be used in various electronic applications including microwave devices, dynamic random access memory (DRAM), non-volatile ferroelectric random access memory (NVFRAM), and next generation gate oxide for IC transistors, to name a few. There has been a renewed interest in developing $Ba_{1-x}Sr_xTiO_3$ and $SrTiO_3$ thin films for frequency agile electronics because the required dc bias for tuning the communication devices can now be drastically decreased because of significant reduction of ferroelectric layer thickness. Epitaxial barium strontium titanate and strontium titanate thin films were deposited on (100) MgO single-crystal substrates CCVD. The as-deposited films were patterned with gold electrodes to form an interdigitated structure, and perform as electrically tunable devices, namely coupled microstripline phase shifters (CMPS). Microwave dielectric properties were characterized at 11-20 GHz with an applied electric field of up to ± 35 V/μm. A maximum figure of merit of 53°/dB at 20 GHz was measured at 23 °C. Capacitance changes of BST films were 3.7% at 35 V and 302 K vs. 3.9 % for PLD films. Dielectric constants of 390-510 were achieved. Thickness variation of BST coatings on 75 mm silicon wafers was less than 2% as measured by ellipsometry. The demonstrated high degree of tunability and relatively low loss showed the great potential of these films for the targeted frequency agile device applications.

2.5. Protective coatings for ceramic matrix composites and corrosion resistance[8–15]

2.5.1. Ceramic matrix composites

So far, ceramic matrix composites (CMCs) have been prevented from fulfilling their significant commercial promise by excessive costs. To place CMCs in a better position as a competitive material it is imperative to develop a fibre coating system that combines the desired materials properties with ease of use and efficiency in a production environment. While alumina fibres have several desirable characteristics for CMC applications, uncoated fibres weaken in the composite due to chemical reactions between the fibre and surrounding matrix. This results in loss of strength and so a thin film fibre coating of a debonding material such as lanthanum phosphate is required to form a reaction barrier. Coatings can also enhance the chemical and abrasion resistance as well as thermal protection of the matrix fibres.

Lanthanum phosphate coatings were deposited onto NEXTEL 610 alumina fibres of up to 0.914 m in length. The coatings are dense and uniform with complete coverage around individual filaments (Fig. 4).

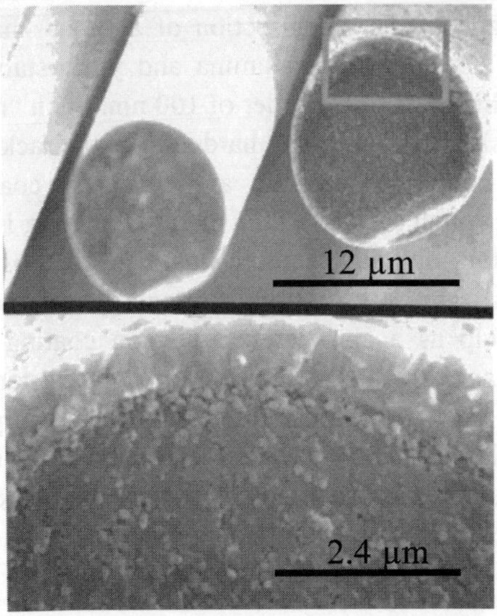

Figure 4. Scanning electron micrograph of 300-400 nm thick NEXTEL alumina fibres coated with lanthanum phosphate. The lower figure is a 5× magnification of the inset in the upper figure.

Process temperatures were as low as 900-1000 °C with only minimal fibre degradation. Beta aluminas, such as $LaAl_{11}O_{19}$ and $BaMg_2Al_{15}O_{27}$, required higher deposition temperatures and did not yield the desired stoichiometry and phases without fibre degradation. The strand strength for coated fibres is 2.72–2.81 kg and is unaffected by thermal ageing. The coatings are phase-pure, columnar grained monazite and vary in thickness from 0.1 to 1.0 μm. Microstructure and compositional layering were uniform among tows comprising 420 and 1260 filaments.

2.5.2. Chromium replacement coatings

The environmental hazards of traditional anti-corrosion coating technologies such as cadmium and chromium plating have motivated considerable R&D to develop environmentally benign alternatives. MCT has prepared several types of coating that could be developed as alternatives to chrome. Figure 5 shows a scanning electron microscope (SEM) image of the cross-section of a CCVD "nanolaminate" ceramic coating, composed of 22 alternating alumina and yttria-stabilized zirconia layers with individual layer thickness of the order of 100 nm. Such "nanostructured" layers promise corrosion resistance, excellent hardness, and crack resistance. Figure 6 shows a proprietary composite ceramic anti-corrosion coating in cross-section (patent pending). The rough surface of the pictured coating is advantageous in, for example, paint adhesion, although smooth coatings are possible with deposition parameter adjustments. This latter coating technology is in advanced development under Navy sponsorship as a chromate conversion coating replacement, having passed relevant ASTM and MIL-SPEC tests for corrosion resistance. This coating has also passed Battelle salt fog testing for 1344 hours, surpassing MIL-C-83488 specifications for a Class I thickness with a chromate conversion coating. These results indicate that CCVD can be used to prepare dense, continuous, pore-free coatings with controlled composition and microstructure.

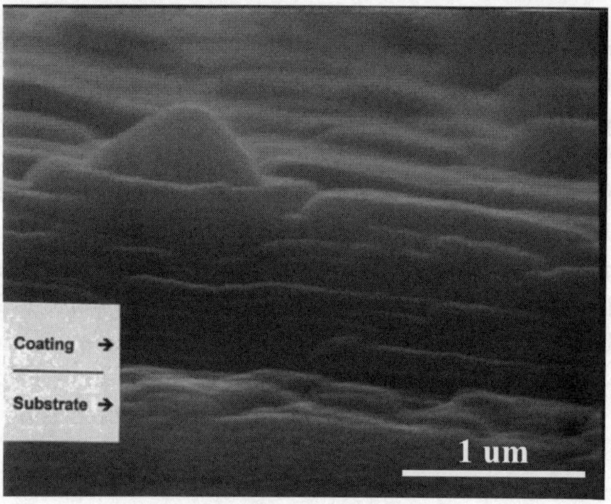

Figure 5. Nanolaminate CCVD coating comprising 22 alternating layers of alumina and zirconia.

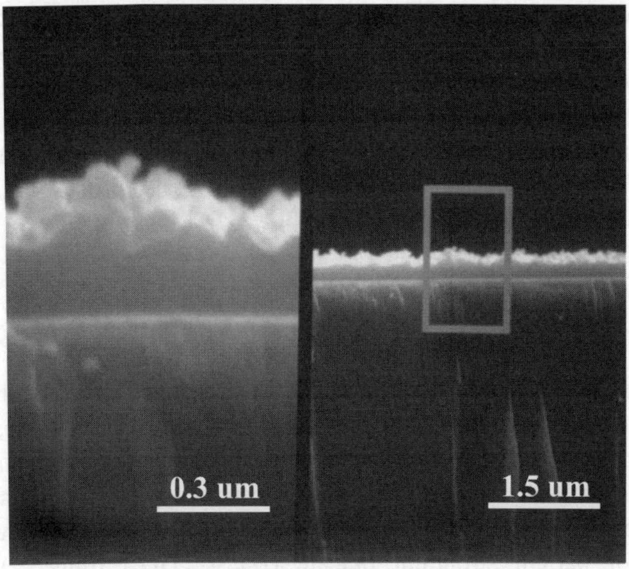

Figure 6. Cross-section of a proprietary corrosion-resistant coating, showing its dense and high surface area nature. The left hand picture is a 5× magnification of the framed area in the right picture.

2.6. Resistor coatings

The CCVD process is well suited to deposit composite resistor layers. The CCVD reagent solutions can be pre-mixed to control the relative amounts of the constituents in each resistor composition (patent pending). Precise control of these ratios permits tailoring of specific resistivity in the resistor layers over many orders of magnitude, ranging from 5 Ω/square to very high sheet resistivity values up to 500 kΩ/square. The resistor material comprises a metal (e.g. platinum or nickel) and a small amount of an oxide (e.g. ceria or silica) added to increase resistivity. There is a linear correlation between the log of resistance and weight percent of a component material in thin film resistors on fused silica substrates. This linearity greatly simplifies the materials engineering to meet the desired resistance specifications. The initial test vehicles have been successfully fabricated with sheet resistivities of 100 Ω/square and 10 kΩ/square. In addition, the process can be used to deposit conducting oxide coatings such as bismuth ruthenium oxide. In amorphous form, these materials have a high resistivity and a positive temperature coefficient of resistance (TCR). In principle, bi-layered coatings consisting of a resistor layer deposited on top of a conductive oxide thin film will result in resistors with zero TCR (patent pending). MCT has already produced high-quality 100 nm resistor coating samples on 69 cm × 83 cm copper sheets and continuous rolled foil, using a multi-flame CCVD system with close to 100% yield. This important preliminary result shows the great promise of the CCVD process in manufacturing embedded passives.

2.7. Flame characterization[16]

In the CCVD process, the flame provides the reactive zone in which precursor materials are pyrolysed and ultimately deposited onto the substrate, hence an in-depth understanding of flame characteristics and processes is a *sine qua non* for rational thin film design. Optical Emission Spectroscopy (OES) and flame pyrometry were used to study the liquid jet flame produced by the Nanomiser device during the deposition of barium strontium titanate ($Ba_xSr_{1-x}TiO_3$) films. While species distribution and vapour-substrate interactions have been studied for a number of chemical vapour deposition (CVD) processes, MCT was interested in correlating flame characteristics with thin film microstructure.

In a series of experiments, the dependence of flame-to-substrate distance and temperatures on the surface morphology of the deposited films was studied. Depositions were performed at 700, 900, 1100 and 1400°C gas temperature (corresponding to distances of 12.2, 11.4, 9.7 and 8.1 cm) respectively. In all cases, the substrate was mounted at a 45° angle with respect to the impinging flame centreline. Optical emission spectroscopy (OES) measurements were performed simultaneously with the flame pyrometry to obtain relative concentrations for emitting species in the flame. Although the observed spectrum provides information about the excited-state density, it does not directly reflect the ground-state population since numerous vapour clusters are formed. OES is a powerful and simple diagnostic tool. The intensity of the OES signal for different species depends on both the species concentration and flame temperature, and can be used to approximate spatial distribution of emitting species.

Both Ba and BaO reach their peak concentration about 1 cm from the nozzle, after which they rapidly decay. The only discrete species emitting in the flame at the distances corresponding to the temperatures studied (8.1 to 12.2 cm) are SrOH and Na. The four films deposited in this temperature range (700 °C to 1400 °C) all yielded X-ray diffraction peaks corresponding to the desired $Ba_xSr_{1-x}TiO_3$ phase. BST films deposited between 700-1100°C were epitaxial; depositions at 1400°C also yielded the desired phase but were not epitaxial.

By combining optical emission spectroscopy and pyrometry, it was possible to derive relative concentrations of short-lived, reactive species in a hydrocarbon flame. Although only relative, these measurements allowed determination of spatial distribution of species within the flame and provided a better description of the flame interaction with the substrate. It has been demonstrated that the excited species emission from the flame reaches a maximum value 1 to 2 cm from the nozzle exit. Determining the relative species concentrations also provided an insight into the mechanism of thin film growth. The observed surface morphology is controlled by the surface mobility of deposited clusters. At low flame and substrate temperatures, the surface mobility is also low, resulting in an even coating with a high surface energy. At higher temperatures, the surface mobility is higher resulting in smoother coatings with lower surface energy. The most important result of this study establishes that precursor pyrolysis, detected by UV-vis spectroscopy, occurs predominately prior to the substrate surface, and high-quality vapour deposited coatings can be produced by this method.

2.8. Fuel cell applications

Fuel cells are of great interest in the technology marketplace, as is illustrated by the sizable investments in the technology and market capitalization of fuel cell companies with little or no current revenues. This is because of their pervasive potential to power homes, offices, shopping malls, automobiles, buses, trains, boats, personal computers, communication equipment and other increasingly distributed, power-demanding appliances of modern living. The overall market size for fuel cells is expected to grow beyond that of transportation and stationary power into distributed 'personal power', to an estimated $1B to $200B annually by 2025. In a very real way, fuel cells will be to energy what integrated circuits have become to information.

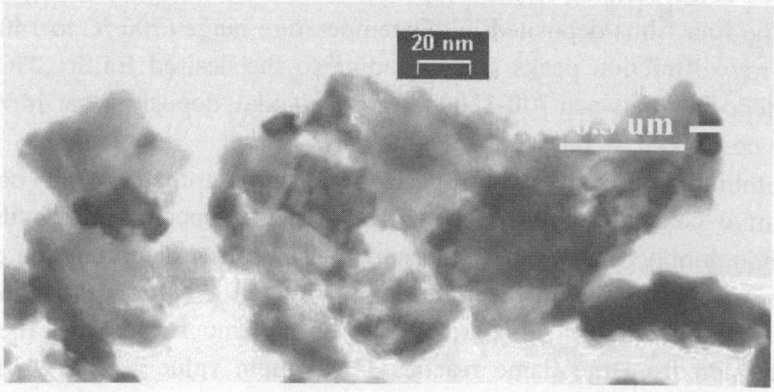

Figure 7. TEM micrograph of platinum nano-particles deposited on to a Nafion membrane. The platinum is dispersed so finely that it appears as a darkening of the Nafion cluster. The darker material is a thicker cluster which was formed when the specimen was abraded from the sample on to the carbon grid for imaging.

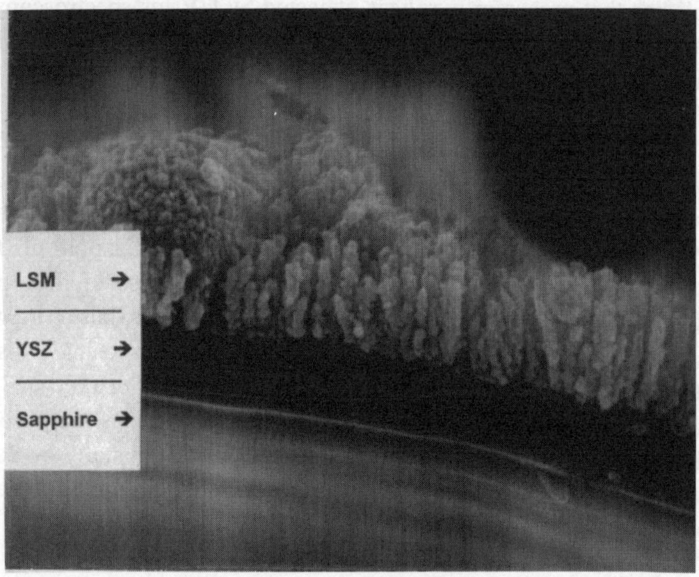

Figure 8. SEM cross-section of a CCVD-deposited LSM conductive coating, showing the open columnar structure over a dense YSZ layer on a sapphire substrate.

For applications in solid oxide fuel cells it is important to increase the interfacial area and total surface area of the coating on the membrane electrode assembly (MEA), while maintaining uniformity. MCT has achieved promising results with the CCVD process to generate advanced catalytic coating structures. Figure 7 shows a transmission electron micrograph of a platinum layer deposited by CCVD on to Nafion: nanoscale platinum particles can be seen as a very fine dispersion of moderate-to-dark grey matter in the light grey membrane material. The scanning electron micrograph in Figure 8 is a cross-sectional view of a conductive lanthanum-strontium-manganate (LSM) layer with a "loose" columnar microstructure deposited on to a dense CCVD layer of yttria-stabilized zirconia. Fully dense YSZ coatings have also been achieved, as have nano-laminate structures.

3. Summary, Conclusion, and Future Directions

The open-atmosphere CCVD technology has many advantages for producing advanced thin film materials when compared with current technologies, in particular it has potential for an order-of-magnitude reduction in operating and capital costs, substantially higher thin film quality, and the ability to engineer materials for specific applications (e.g. excellent compositional and structural control). In addition CCVD is easy to use which permits both rapid development of new coating/substrate combinations and straightforward implementation for production-friendly manufacturing. Finally, the CCVD process, in conjunction with the Nanomiser device, uses only environmentally benign chemical precursors, minimizing safety and pollution concerns.

References

[1] Hunt A.T., Carter W.B, Cochran J.K., *Method and Apparatus for the Combustion Chemical Vapor Deposition of Films and Coatings* (1997): U.S. Patent 5,652,021

[2] Lin W.-Y., *Embedded Capacitors for Electronic Circuits Based on Flexible Laminate Multichip Module/Printed Circuit Boards* (1998) DARPA STTR Phase I grant: DAAHO1-98-C-R033

[3] Hunt A.T., *CCVD Processing of Lanthanum Phosphate and Beta-Alumina Fiber Coatings for Mullite Matrix Composites* (1996) NSF SBIR Phase I grant: DMI9561712, 1996.

[4] a) Lin W.Y, *Low Loss and Cost Thin Film for Frequency Adaptive Electronics* (1998) DARPA SBIR Phase I grant: DAAHO1-97-C-R230 (b) Paranthaman, M., *Private Communication* (c) Shoup S.S., *Buffer Layers on Textured Nickel Using Commercially Viable CCVD Processing* (1997) DOE SBIR Phase I grant: DE-FG02-97ER82345 (d) Shoup S.S., *et al*, *IEEE Transactions on Applied Superconductivity* (submitted)

[5] Lin W.Y., Speyer R.F., Shrout T.R., Hackenberger W.S., *J. Am. Ceram. Soc.*, **5** (1999) 1207

[6] Hunt A.T., Hörnis H.G., *CCVD Theory and Processing of Ferroelectric Thin Films at Ambient Pressure* (1995) MOCVD Workshop for Silicon Processing, Kyungju, Korea

[7] Hunt A. T., Abbey J., Hörnis H. G., *PZT, PLZT, LSC and Pt Thin Films Produced in the Open Atmosphere Using Combustion Chemical Vapor Deposition* (1995) International Symposium on Integrated Ferroelectrics, Colorado Springs, CO

[8] (a) Hendrick M.R., *The Effects of Combustion CVD-Applied Alumina Coatings on the High Temperature Oxidation of a Ni-Cr Alloy* (1996) Georgia Institute of Technology, Atlanta (Thesis) (b) Lin W.Y., *et al*, *Epitaxial* $Ba_{1-x}Sr_xTiO_3$ *and* $SrTiO_3$ *Thin Films for Frequency Agile Wireless Communications* (1998) 31st International Symposium on Microelectronics in conjunction with The American Ceramic Society Electronics Division Fall Meeting, San Diego, CA

[9] (a) Hendrick M. R., Shao H., Hwang T. J., Hunt A. T., *Low Cost Corrosion Protection of Metals by Combustion Chemical Vapor Deposited Coatings* (1998) Materials Solutions Conference, Rosemont, Illinois (b) Lin W.Y., *et al*, *Thin Film Dielectrics for Electronics Using Combustion Chemical Vapor Deposition* (1999) 26th Spring Meeting of the Materials Research Society, San Francisco

[10] Hendrick M.R., Hampikian J.M., Carter W.B., *J. Electrochem. Soc.,* **145** (1998) 3986

[11] Hendrick M.R., Shanmugham S., Hunt A.T., *Elevated Temperature Oxidation Protection of Carbon Steel by Combustion Chemical Vapor Deposition* (1999) Symposium on High Temperature Coatings III, TMS Annual Meeting, San Diego, CA.

[12] Hendrick M.R., Hampikian J.M., Carter W.B., *Combustion Chemical Vapor Deposition of Alumina* (1995) Gordon Research Conference on High-Temperature Corrosion, New London, New Hampshire

[13] Hendrick M.R., Shanmugham S., Hunt A.T, *Chrome Conversion Replacement Coatings via CCVD Processing* (1998) US Navy and Industry Corrosion Technology Information Exchange, Louisville, Kentucky

[14] Hendrick M.R., *et al*, *Chrome Replacement and Thermal Corrosion Resistant Coatings Via Open Atmosphere CCVD Processing* (1996) 3rd Annual Advanced Techniques for Replacing Chromium, Seven Springs, Pennsylvania

[15] Hunt A. T., Hendrick M.R., Shao H., McEntyre J. E., Hwang T. J., *Chrome Replacement Coating Methods and the Potential Use of CCVD Processin* (1997) ASM Materials Solutions Conference, Indianapolis

[16] Oljaca M., Luten H.L., Tomov T., Metzger T., *Combust. Flame* (submitted)

Additional publications about Combustion CVD by our scientists and collaborators:

[17] Carter W.B., Book G. B., Hunt A.T., *Applied Superconductivity*, (1993) 391

[18] Carter W. B., Hampikian J. M., Godfrey S., Polley T. A., *Mater. Manufacturing Processes*, **10** (1995) 1007

[19] Hunt A.T., Carter W.B., Cochran J.K. Jr., *Appl. Phys. Lett.*, **63** (1993) 266

[20] Hunt A.T., *Combustion Chemical Vapor Deposition from Liquid Organic Solutions* (1993) Georgia Institute of Technology (Thesis)

We are aware of recent work by others concerning development of flame-based thin film deposition processes. However, our efforts and patents predate all of this work. Some key references:

[21] Choy K.L., Wiedman I., Derby B., *Adv. Sci. Technol.*, 5 (1995) 299

[22] Choy K.L., *British Ceramic Proceedings*, 54 (1997) 65

[23] Choy, K.L., Charojrochkul S., Steele B.C.H., *Solid State Ionics*, **96** (1997) 49

[24] Skandan G. *et al*, *J. Am. Ceram. Soc.*, **81**(1998) 2753

[25] Nanopowder Enterprises, (1997) Inc., *Low-Temperature Processing of Thick Film Ferroelectric Materials,* STTR (BMDO): Abstract
[26] Glumac N.G., Chen Y.J., Skandan G., *J. Mater. Res.*, 13 (1998) 2572

CHAPTER SIX

HYPERSONIC PLASMA PARTICLE DEPOSITION OF NANOCRYSTALLINE COATINGS

S. L. Girshick[a], J. V. R. Heberlein[a], P. H. McMurry[a], W. W. Gerberich[b], D. I. Iordanoglou[a], N. P. Rao[c], A. Gidwani[a], N. Tymiak[b], F. Di Fonzo[a], M. H. Fan[a], and D. Neumann[a]

[a]Department of Mechanical Engineering
[b]Department of Chemical Engineering and Materials Science
University of Minnesota, Minneapolis, Minnesota, USA

[c]Microtherm, LLC, Minneapolis, Minnesota, USA

1. Introduction

The mechanical properties of nanostructured materials may be significantly enhanced in comparison with those of coarser-grained materials with the same chemical composition. Potentially dramatic improvements can be achieved in hardness, yield strength, and sintering and densification behavior [1-10] making nanostructured materials attractive candidates for advanced friction- and wear-resistant coatings. The wear resistance of hard but brittle materials such as ceramics can be significantly improved by increasing fracture toughness and by decreasing modulus, friction coefficient and inherent flaw size [11], and nanostructured materials are expected to produce a favourable combination of these properties.

Thermal spraying with plasmas or flames is a widespread, highly developed technology for applying coatings for wear resistance as well as for other functions such as corrosion resistance and as thermal barriers. In thermal spray processes a relatively coarse powder (20-100 μm diameter) is injected into a spray torch. The individual particles are melted by the hot gas as they accelerate towards the substrate and on impact these molten particles form splats, thereby building up a continuous coating. A crucial advantage of thermal spray technology is that the deposition rates, of the order of cm/hr, are orders of magnitude higher than those in chemical vapor deposition (CVD) processes, however the coating microstructure is generally coarse, having essentially the same grain size as that of the powder injected into the torch. Several research groups are currently studying the use of conventional thermal spray equipment to deposit coatings with various degrees of nanocrystallinity, either by injecting coarse powder that has been pretreated to induce nanocrystallinity [12,13], by injecting atomized droplets containing nonvolatile components in solution [14], or by high-velocity particle impact [15].

In contrast to these approaches, our group at the University of Minnesota has developed a process, hypersonic plasma particle deposition (HPPD) [16], in which a nanostructured coating is deposited by spraying nanoparticles themselves, but no powder is injected into the torch and only vapour-phase reactants are introduced into the plasma, which is then expanded through a nozzle. Rapid cooling in the nozzle drives the nucleation of particles, which are then accelerated to high enough velocities in the hypersonic flow downstream of the nozzle to ensure that particles larger than a few nanometers in diameter deposit by inertial impaction. For example, our results indicate that for typical conditions involving synthesis of silicon carbide, the mean particle size is about 20nm and the particle impact velocity is about 1700 m/s, resulting in a dense coating with a grain size of about 20 nm.

The overall experimental apparatus used for HPPD is described in Section 2 below. Nanoparticle synthesis in thermal plasmas and subsonic expansions are reviewed in Section 3. In Sec. 4 the rationale for the HPPD process is presented and in Section 5 the results for nanoparticle synthesis in a supersonically expanded plasma. In Section 6 the coating formation process is discussed and the results of film characterization reviewed. Preliminary studies of post-processing the film are discussed in Section 7 and finally in Section 8 a recently developed modification of HPPD is discussed, in which the deposition substrate is replaced by a system of aerodynamic lenses, which focuses the nanoparticles to a collimated beam with a width of tens of microns. This beam is used to "write" lines and two-dimensional patterns on a translatable substrate, thereby fabricating nanocrystalline microstructures.

2. Experimental Apparatus

The apparatus used for the HPPD experiments is shown in Fig. 1. The plasma is generated by a direct current (DC) arc operating at about 225 A current and 10 kW power. The main plasma gases are argon and hydrogen, with typical flow rates of 30 slm and 7.5 slm respectively. Gaseous reactants for particle synthesis are injected radially through an injection ring located at the upstream end of a boron nitride nozzle. The DC torch, injection ring and nozzle have separate water cooling circuits, which allows for calorimetric determination of the enthalpy flow rate at several axial locations. The pressure is 60–70 kPa at the nozzle inlet and 265–350 Pa in the deposition chamber (apart from the subsonic flow experiments, see below). The nozzle is 100 mm long (50 mm for the subsonic flow experiments) with an entrance diameter of 15 mm and an exit diameter of 5 mm.

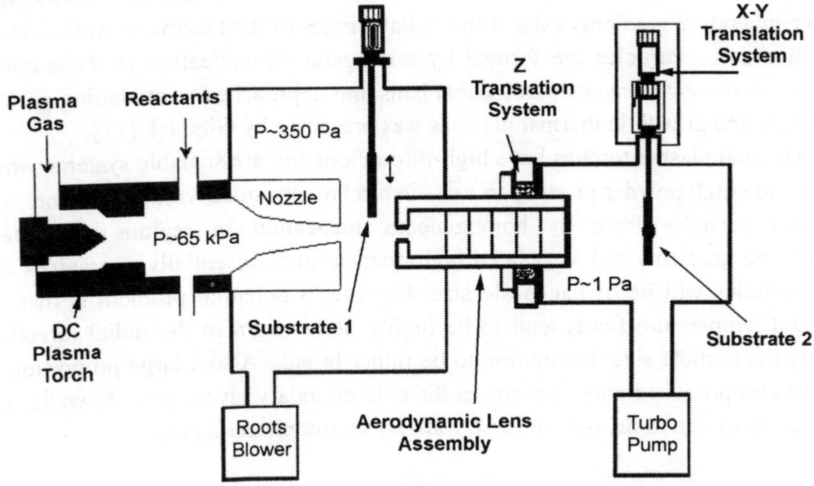

Fig. 1. Overall schematic of experimental apparatus. "Substrate 1" is used for deposition of continuous films. "Substrate 2" is used for focused particle beam deposition of patterned microstructures.

For depositing continuous films a substrate was positioned 20 mm downstream of the nozzle exit. In some cases this substrate was replaced by an aerosol sampling probe. For the focused particle beam experiments the substrate was removed, and an aerodynamic lens assembly was placed at the downstream end of the deposition chamber. The particle beam exiting the lens assembly emerged into a second chamber, at a background pressure of 1.0 Pa, and particles were deposited on a translatable substrate.

Further details of the apparatus and operating conditions are discussed in the following sections.

3. Particle Synthesis

3.1. Thermal plasma synthesis

The use of thermal plasmas for ultrafine powder synthesis has been extensively investigated. These types of plasma are in the atmospheric-pressure regime, at least in the plasma generation region, and are generated either by DC arcs or by radiofrequency (RF) induction. Peak temperatures are quite high, of the order of 10,000 K, so that reactants injected in gaseous form can be almost completely dissociated. Steep temperature gradients exist at the radial fringes of the plasma as well as in the flow direction. Particles are formed by homogeneous nucleation in these cooler regions. A review of nucleation mechanisms and approaches to modelling particle nucleation and growth in thermal plasmas was presented by Girshick [17].

Thermal plasma torches have high-throughput and are scalable systems which can provide high powder production rates in a relatively small-volume reactor. The fact that particles form by homogeneous nucleation in regions with steep temperature gradients and associated high quench rates, essentially guarantees that these particles will be of nanoscale size, however a potential problem is that the flow and temperature fields tend to be highly nonuniform in the radial direction, causing the particle size distribution to be rather broad. Also a large proportion of the particles produced may nucleate in the cold boundary layers close to walls, and will then be driven to deposit on these walls by thermophoresis [18].

3.2. Particle synthesis with a thermal plasma expanded through a subsonic nozzle

To overcome these difficulties we developed the approach of expanding the plasma to lower pressure through a hot-walled nozzle, and causing particle nucleation to occur within the nozzle. Nozzle expansions provide a well-characterized environment in which the temperature drops rapidly in the flow direction, while the accelerating flow is relatively uniform in the radial direction, apart from a thin boundary layer near the wall. By combining a thermal plasma source with a nozzle expansion, one can generate nanoparticles at high rates, with relatively narrow size distribution.

Initial experiments [19,20] involved the synthesis of silicon, carbon and silicon carbide particles using a subsonic expansion. The reactants consisted of vapour-phase silicon tetrachloride and/or methane, according to whether the desired product was Si, C or SiC. The silicon tetrachloride flow rate was in the range of 50–70 sccm, while the methane flow rate ranged from 100 to 1000 sccm. Other operating conditions were similar to those discussed in Section 2 above, except that the pressure was typically about 67 kPa at the nozzle inlet, and 53 kPa at the nozzle exit. Calorimetric measurements, together with a quasi-one-dimensional flow model which accounted for finite-rate hydrogen recombination [21], indicated that for these basic conditions the mean temperature and velocity at the nozzle exit were approximately 2100 K and 430 m/s, respectively.

The aerosol issuing from the nozzle was characterized by inserting a capillary sampling probe into the reactor, followed by a two-stage ejection diluter to dilute the sampled gas by approximately a factor of 1000 in dry filtered argon, in order to quench coagulation rapidly. The sampled stream was then directed to two separate instruments: a scanning electrical mobility spectrometer (SEMS) [22] and a small hypersonic impactor [23]. The SEMS, which consists of a differential mobility analyser [24] in series with an ultrafine condensation particle counter [25], provides on-line particle size distributions which are accurate down to about 4 nm in diameter. The hypersonic impactor deposited representative samples onto transmission electron microscopy (TEM) grids for later microscopy and other analysis.

Fig. 2 shows representative particle size distributions measured with the sampling probe aligned coaxially with the flow and positioned 25 mm downstream of the nozzle exit. Repeated experiments showed good reproducibility for given conditions. For both the silicon and silicon carbide particles the size distributions have well defined peaks at about 10 nm, with about 90% of the detected particles being smaller than 16 nm in diameter. For the carbon particles on the other hand, the trend in the data indicates that a large fraction of the particles were smaller than the 4-nm detection limit of the SEMS. Indeed, carbon particles could only be detected if the hydrogen flow was omitted. With hydrogen (at the same flow rate as for the Si and SiC experiments) plus methane, no particles were detected above the background level.

These results, together with numerical modelling of flow and aerosol dynamics in both the nozzle and the free jet expansion downstream of the nozzle [20,26], and the observation that the molar volume of silicon carbide is only 3% larger than that of silicon, lead to the conclusion that silicon carbide particles under these conditions are formed by homogeneous nucleation of small silicon clusters from supersaturated silicon vapor, followed by their carburisation by surface reactions. For the same conditions as in the silicon and silicon carbide experiments in Fig. 2, the model predicts that the number-mean particle diameter at the nozzle exit is about 2 nm, and that coagulation in the free jet (and, to a small extent, in the sampling capillary) causes the particles to grow to the size measured by the SEMS.

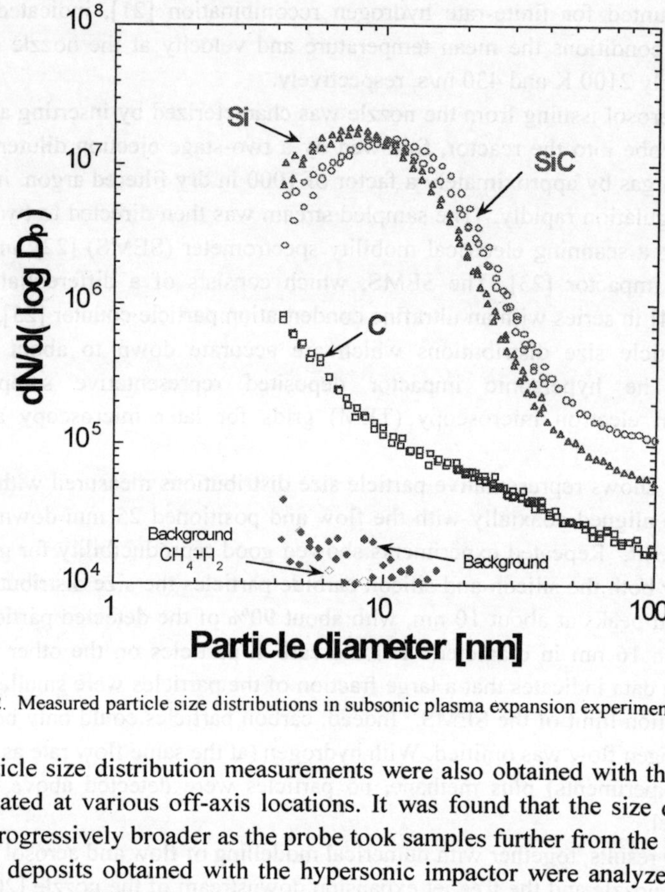

Fig. 2. Measured particle size distributions in subsonic plasma expansion experiments [19].

Particle size distribution measurements were also obtained with the sampling probe located at various off-axis locations. It was found that the size distribution became progressively broader as the probe took samples further from the flow axis.

The deposits obtained with the hypersonic impactor were analyzed by TEM and by electron diffraction [19]. The silicon agglomerates tended to be more compact, showing a tendency to coalesce, whereas the silicon carbide particles tended to retain their primary particle size, and appeared to be joined together in aggregates by partial sintering. This result, which is well known in the literature for many ceramic materials produced in high temperature synthesis processes such as plasmas or flames, points to a potential problem because partially sintered aggregates are deleterious for subsequent densification processes.

4. Rationale for Operating in the Hypersonic Flow Regime

The subsonic expansion experiments discussed above demonstrated that a thermal plasma expanded through a nozzle is an efficient means of generating particles of about 10 nm in diameter, but they also highlighted some potential problems. One problem is that the long residence times in the downstream chamber allow considerable time for coagulation, leading materials such as silicon carbide to form low-density aggregates. A second problem is that although the flow is close to one-dimensional inside the nozzle, it is radially nonuniform in the free jet expansion downstream of the nozzle and in fact flow recirculation may take place in the downstream chamber, so that the particle size distributions become rather broad as one moves away from the flow axis. A further issue is that the high specific surface area of these powders renders them extremely prone to contamination.

In principle HPPD overcomes these difficulties. First, the problem of in-flight agglomeration is minimized by operating in highly supersonic mode, because the residence time available for coagulation downstream of the nozzle is sharply reduced. Based on the measured particle concentrations accounting for the sampling dilution ratio of approximately 10^{10} cm^{-3}, one can predict [27] that the characteristic time for coagulation (time required for the particle concentration to decrease by a factor of two) is approximately 0.1 s. Thus for a subsonic flow issuing into a large chamber ample time is available for coagulation, as was clearly shown by measuring size distributions at different axial locations [20]. For the hypersonic flow used for HPPD, however, the residence time for a 20 nm particle between the nozzle exit and the deposition substrate is only about 10 μs (an average velocity of about 2000 m/s over a distance of 20 mm), essentially eliminating coagulation downstream of the nozzle.

The absence of coagulation between the nozzle and the substrate means that the particle size distribution does not vary significantly in the radial direction, at least not in the short distance of about four nozzle exit diameters between the nozzle exit and the substrate, although there may be some radial variation in the particle flux over the substrate.

Finally, the contamination problems associated with a loose ultrafine powder are mitigated by placing a deposition substrate downstream of the nozzle, assuming that the deposit is reasonably dense. The density of the deposit depends on the deposition mechanism. Deposition modes can be categorized as involving either "soft landing" or "ballistic impact". "Soft landing" mechanisms include gravitational settling, Brownian diffusion and thermophoresis. Gravitational settling is negligible for the particle sizes and time scales involved. Brownian diffusion and thermophoresis are not necessarily negligible for the pertinent conditions, but the associated impact velocities are relatively small. "Ballistic impact" refers to inertial deposition, which for nanoparticles is possible only if the particle velocity is quite high. In that case some degree of densification may be achieved by ballistic consolidation.

Hypersonic impaction of nanoparticles was previously studied theoretically and experimentally by Fernandez de la Mora and coworkers [23,28,29], who showed that one can deposit particles as small as a few nm in diameter from hypersonic free jets with expansion ratios of the order of several hundred. Particles above a well-defined cut size deposit with nearly 100% efficiency, while particles below the cut size have insufficient inertia to penetrate the flow streamlines as they curve round the substrate. The distance between the nozzle exit and the substrate controls the extent of the expansion and is thus a crucial parameter in determining the impaction cut size.

5. Particle Synthesis with a Supersonically Expanded Plasma

As noted in Section 4.2 above, the mean particle diameter at the nozzle exit in the subsonic expansion experiments was estimated to be about 2 nm. For our range of film deposition conditions we calculate the cut size for inertial deposition to be in the range 3 to 6 nm. Indeed, we believe this to be the explanation for the fact that in our first attempt at HPPD we failed to obtain a deposit because the particles, having insufficient time to grow by either coagulation or surface reactions, were too small to impact. To solve this problem a 5-mm-long extension was added to the nozzle that had been used in the subsonic flow experiments, doubling its total length, to allow time for particle growth to occur within the nozzle. (Note that the nozzle was not adiabatic. The upstream half was mounted in a water-cooled holder, while the extension lost heat by radiation to the cold chamber walls.)

The particle size distributions were then characterized by the SEMS for supersonic flow conditions [30], with the sampling probe placed at the same location, 20 mm downstream of the nozzle exit, as the typical deposition substrate position for the HPPD experiments discussed below. Fig. 3 shows measured particle size distributions for downstream chamber pressures of 100, 50 and 20 Torr (13.3, 6.7 and 2.7 kPa), all of which lay in the supersonic regime. The peak in the distribution shifts toward smaller sizes with reduction in pressure, although this effect is relatively small with the peak in all cases found to be around 20 nm. A more obvious effect as the pressure is reduced however, is that the size distribution becomes narrower, presumably as a consequence, of the increased velocity and decreased time for coagulation. At 20 Torr the particle size distribution is closely fit by a lognormal distribution with a geometric standard deviation of 1.55, approaching the theoretically predicted limiting value of 1.46 for a free-molecule self-preserving size distribution [31].

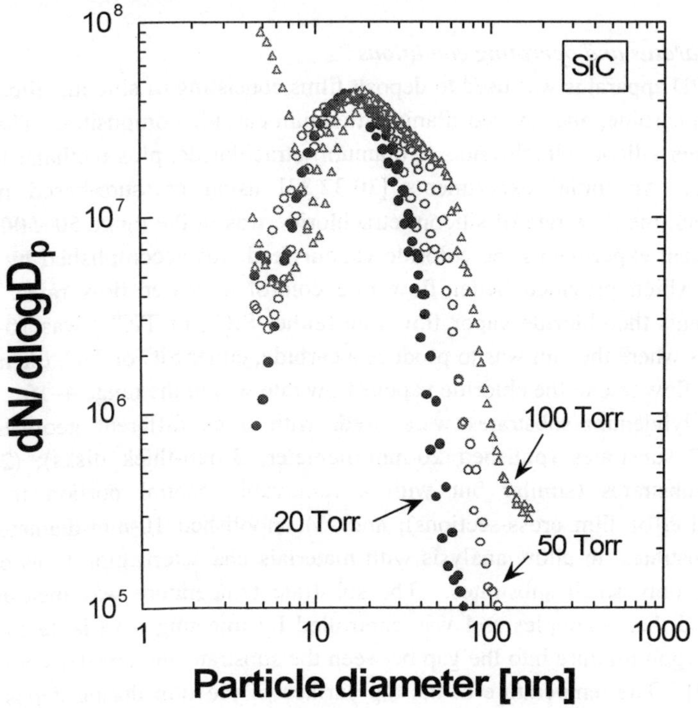

Fig. 3. Measured particle size distributions in supersonic plasma expansion experiments [30].

For the film deposition experiments discussed in the next section the chamber pressure was in the range 2.0–2.6 Torr (265–350 Pa). Unfortunately, for the SEMS measurements we were unable to sample below 20 Torr, because of pumping limitations on the ejection dilution system. The two-stage ejection diluter boosts the atmospheric pressure to the value at which the differential mobility analyzer is designed to operate, however the trend shown in Fig. 3 clearly suggests that the particle size distribution should be even narrower under film deposition conditions than in the 20 Torr measurements, with a mean diameter of about 20 nm, assuming the same upstream conditions.

6. Film Deposition

6.1. Apparatus and operating conditions

The HPPD apparatus was used to deposit films consisting of silicon, silicon carbide, titantium carbide, and layered titanium/titanium carbide composites. The reactants were either silicon tetrachloride or titanium tetrachloride, plus methane for carbide synthesis. In initial experiments [30,32,33] using pressure-based mass flow controllers, the flow rate of silicon tetrachloride was in the range 50–500 sccm. In more recent experiments the chloride vapour feed was accomplished by a bubbler system, which provided better flow rate control at lower flow rates. In these experiments the chloride vapor flow rate (either $SiCl_4$ or $TiCl_4$) was 10–40 sccm. For cases where the aim was to produce a carbide, either SiC or TiC, the ratio of the methane flow rate to the chloride vapour flow rate was in the range 4–10.

Molybdenum substrates were used, with three different geometries: (1) "normal" substrates (polished 20-mm-diameter, 3-mm-thick disks); (2) "break-apart" substrates (similar but with a removable central portion to facilitate observation of film cross-sections); and (3) unpolished 10-mm-diameter, 4-mm-thick substrates, to allow analysis with materials characterization tools capable of handling only small substrates. The substrate temperature was measured using embedded thermocouples and was controlled by injecting a variable-composition helium-argon mixture into the gap between the substrate and a water-cooled copper base [34]. The temperature at the top surface of the film during deposition may have been somewhat higher than the reported thermocouple temperature.

6.2. Numerical modelling of flow and particle trajectories

A two-dimensional (axisymmetric) numerical model was developed to simulate the compressible flow in the nozzle and in the free expansion region between the nozzle and the substrate, for the same geometry and operating conditions as in the experiments. The flow was assumed to consist of Ar, H and H_2, in chemical equilibrium at the inlet. Finite-rate hydrogen recombination was modelled, including both three-body reactions in the gas phase and heterogeneous reactions at the nozzle wall. The model was tuned to match the measured convective heat transfer from the nozzle to the water-cooled nickel holder for the first half of the nozzle length, and accounted for radiative heat transfer from the nozzle to the cold chamber walls for the portion which protruded past the nickel holder. The fluid conservation equations were solved using the code CFD-ACE [35], with the chemical species equations covered by the CHEMKIN suite of codes [36].

Fig. 4 shows the predicted isocontours of the gas axial velocity in the nozzle and in the free jet expansion. For these calculations the temperature at the nozzle inlet was set to 4000 K, a representative value based on calorimetric measurements. The substrate temperature was set to 600 K, the argon and hydrogen flow rates were respectively 30 slm and 7.5 slm, and the background pressure downstream of the nozzle was set to 267 Pa. As can be seen from the velocity contours, a bow shock is predicted to occur 2.5 mm (on the centerline) in front of the substrate. The location and shape of the shock are in excellent agreement with an analysis of photographs taken during the experiments using a colour video camera with a telephoto zoom lens [37]. The axial velocity is predicted to peak at the shock at a value of approximately 2400 m/s, and then to drop sharply. The peak Mach number is predicted to equal 7.1. The minimum temperature in the expansion is predicted to be 193 K, recovering to almost 4000 K immediately downstream of the shock and then dropping sharply to the substrate temperature. It should be noted that this environment may be conducive to film growth on the substrate by chemical vapour deposition (CVD). Thus, while evidence clearly indicates that film deposition is mainly by impacting particles, it is possible that film growth by CVD occurs simultaneously, and that the extent to it does so would be expected to be sensitive to substrate temperature.

Particle trajectories were modelled on the assumption that the gas velocity and temperature between the nozzle and the substrate are not affected by the presence of particles. Thus the flow field solution was used as a starting point for the particle trajectory calculations, which used a Lagrangian approach. Particles were assumed to be spherical, and to exit the nozzle at various radial locations with the same velocity and temperature as the gas at that location. The model accounted for viscous drag and convective heat transfer between the gas and the particle.

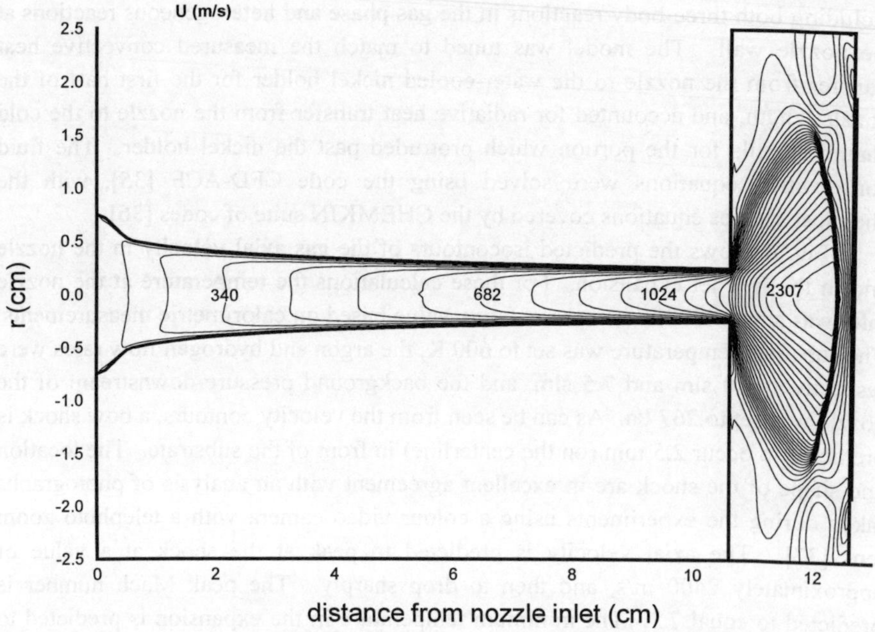

Fig. 4. Predicted isocontours of gas axial velocity (m/s) in nozzle and free expansion, terminating at deposition substrate on right.

Fig. 5 shows predicted trajectories for silicon carbide particles of various diameters, under the same conditions as in Fig. 4. The trajectories shown are for particles which exit the nozzle at a radial location midway between the centreline and the nozzle wall. The greater inertia of larger particles keeps them closer to the centreline, but even 5-nm particles are predicted to impact on the substrate, whereas 1-nm particles are predicted to follow the flow as it curves round the substrate. The predicted cut size here is around 4 nm.

Predicted particle velocities are shown in Fig. 6. The particle size determines the degree to which the particle is able to follow the flow as it accelerates in the free expansion downstream of the nozzle and then sharply decelerates immediately after the bow shock. For 20-nm particles the predicted impact velocity is approximately 1700 m/s.

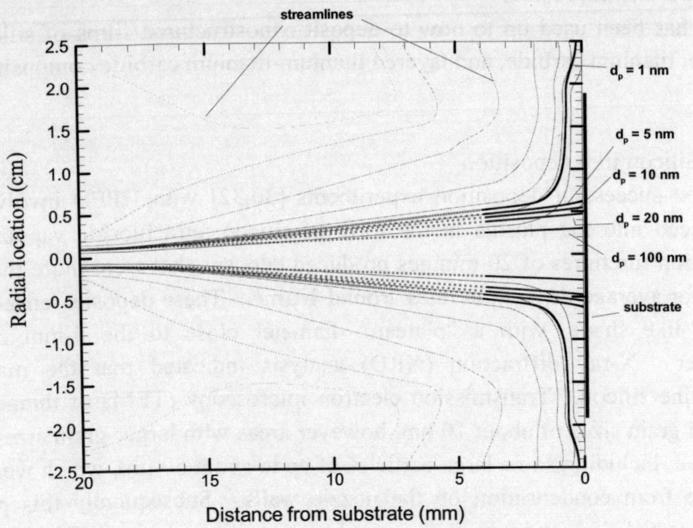

Fig. 5. Predicted gas streamlines and particle trajectories between the nozzle exit and the substrate, for silicon carbide particles of various diameters d_p.

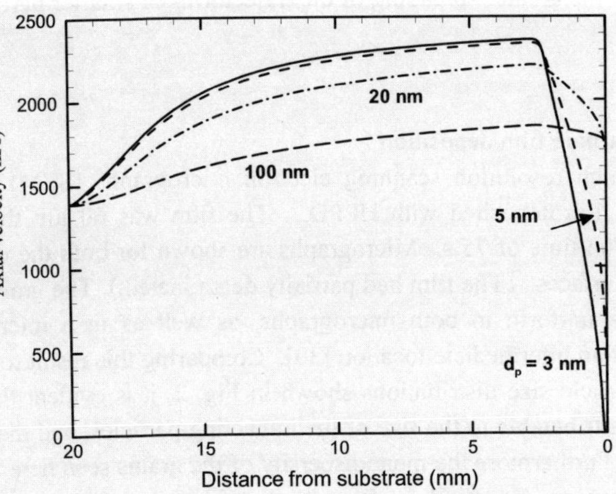

Fig. 6. Predicted axial velocities of silicon particles of various diameters, for same conditions as Fig. 5.

6.3. Experimental results

HPPD has been used up to now to deposit nanostructured films of silicon, silicon carbide, titanium carbide, and layered titanium-titanium carbide composite.

6.3.1. Silicon film deposition

The first successful deposition experiments [30,32] with HPPD involved silicon, introduced into the plasma in the form of silicon tetrachloride vapour. Typical deposition run times of 20 minutes produced deposits that were more than one mm thick, for average deposition rates around 1 μm/s. These deposits tended to have a "mesa"-like shape, with a "plateau" diameter close to the 5-mm nozzle exit diameter. X-ray diffraction (XRD) analysis indicated that the material was crystalline silicon. Transmission electron microscopy (TEM) of thinned samples showed grain sizes of about 10 nm, however areas with larger grain sizes were also observed, including some large particles of up to about 50 μm, which were believed to arise from condensation on the nozzle walls. Subsequently this problem of "macroparticle" ejection from the nozzle was mitigated by modifying the nozzle design to run with hotter walls [21], by keeping run times short, and/or by running with lower reactant flow rates. With macroparticles mostly if not entirely eliminated, deposits tended to coat the entire 20-mm-diameter substrate. This may be explained by the greater spreading of the trajectories of smaller particles, as seen in Fig. 5.

6.3.2. Silicon carbide film deposition

Fig. 7 shows high resolution scanning electron micrographs (SEM) of a typical silicon carbide film deposited with HPPD. The film was 60 μm thick and was deposited in a run time of 75 s. Micrographs are shown for both the top (left) and bottom (right) surfaces. (The film had partially delaminated.) The grain size, about 20 nm, appears uniform in both micrographs, as well as in a micrograph (not shown) taken of an intermediate location [30]. Comparing this result to the ~20-nm peaks in the particle size distributions shown in Fig. 2, it is evident that the grain size is directly attributable to the size of the impacting particles, and that little grain growth occurs. Furthermore the monodispersity of the grains seen here is consistent with the observation discussed in Sec. 4, that lowering the chamber pressure reduces the width of the particle size distribution, the chamber pressure here being 333 Pa.

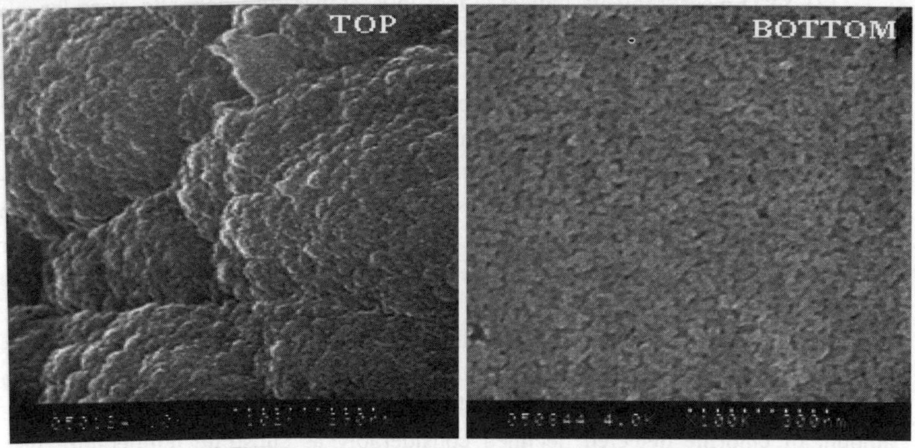

Fig. 7. Scanning electron micrographs of the top and bottom surfaces of a silicon carbide film. Film thickness was 60 μm, deposited in a run time of 75 s.

The effect of substrate temperature on film properties was investigated by depositing silicon carbide films with substrate temperatures ranging from 250°C to 750°C [33]. These experiments indicated that substrate temperature did not affect grain size, again suggesting that grain growth during deposition was minimal. This result is consistent with film post-treatment experiments, discussed in Sec. 7 below, in which no grain growth was observed until the temperature was somewhere in the range 900–1000°C. However several other film properties *were* affected by substrate temperature, including the following.

(1) Under routine transport and handling, films deposited at low substrate temperatures tended to delaminate easily, whereas films deposited at high substrate temperatures were more adherent. A microindenter with a Vickers indenter tip was used to test the adhesion of two films of equal thickness, one deposited at a substrate temperature of 540°C, the other at 700°C. With an indentation load of 1 kg, the lower-temperature deposit underwent extensive spalling, while the higher-temperature deposit showed only some cracking and partial delamination. In order to obtain quantitative adhesion measurements, a 1-μm tungsten overlayer was sputtered over films deposited at various substrate temperatures. For a low-temperature deposit, measurements obtained from delamination induced by a 900-mN load indicate an adhesion strength of 2.7–3.9 J/m^2. No delamination was observed for the high-temperature film.

(2) XRD measurements indicated that low-temperature deposits were largely amorphous, while high-temperature deposits showed the presence of well-defined crystalline phases, primarily β-SiC and Mo_2C. Rutherford backscattering spectrometry (RBS) detected no elemental molybdenum within the penetration depth (~6 μm) of the probe beam into the top surface of these relatively thick (>50 μm) films, suggesting that the molybdenum carbide formed as a layer at the film-substrate interface. This interlayer, presumably formed by interdiffusion between the molybdenum substrate and carbon atoms from either impacting particles or adsorbed gas-phase species, would be expected to enhance adhesion through chemical bonding.

(3) RBS measurements indicated that the chemical composition of the films was close to stoichiometric SiC at all substrate temperatures, however measurable chlorine contamination was found in all cases, perhaps adsorbed as HCl, and the amount of chlorine detected declined considerably as substrate temperature was increased, from 17.6 mole% in a film deposited at a substrate temperature of 250°C, to 0.5% in a film deposited at 650°C. Post-deposition experiments were conducted in which the film was exposed briefly to a high-temperature RF plasma, and they showed that chlorine contamination can be further reduced by post-deposition heat treatment.

(4) As-deposited film density increased as substrate temperature increased. Based on the absolute elemental concentrations in the films measured by RBS, the film density ranged from 65% of the theoretical density (3.21 g/cm^3) of SiC for a substrate temperature of 250°C to 95% of theoretical density for films deposited at 650–700°C. This result may be due to the effect of substrate temperature on the deformation behaviour of the particles upon impact, to the simultaneous occurrence of chemical vapor deposition film growth, or to a combination of these.

(5) Film hardness increased with increasing substrate temperature. Hardness and Young's modulus were measured using two nanoindentation instruments, a Hysitron Triboscope operating in conjunction with an atomic force microscope, and a micromechanical tester (MMT). The Hysitron provides greater accuracy at small loads, up to a few mN, while the MMT can provide loads up to 900 mN. Nanoindentation tests were performed using conical 90-degree indenters of radius of 0.4 μm for the Hysitron and 1.0 μm for the MMT. Hardness evaluation was based on load-displacement curve analysis [38]. Fig. 8 shows results for two films, one deposited at a substrate temperature of 450°C, the other at 700°C, with film thicknesses of 6–10 μm. Each data point is an average based on several load-displacement curves. The rise and fall of each curve with indenter penetration depth is as

expected, as measurements closest to the surface are artificially low due to surface roughness, while measurements past the peak in the curves are affected by the presence of the relatively soft molybdenum substrate. In any case, it is evident that the higher-temperature deposit was significantly harder. This effect is presumably related to the increase in as-deposited film density with substrate temperature. The peak in the hardness measurements for the film deposited at 700°C is at 37.3 GPa, above the 26–36 GPa range of standard bulk silicon carbide [39,40].

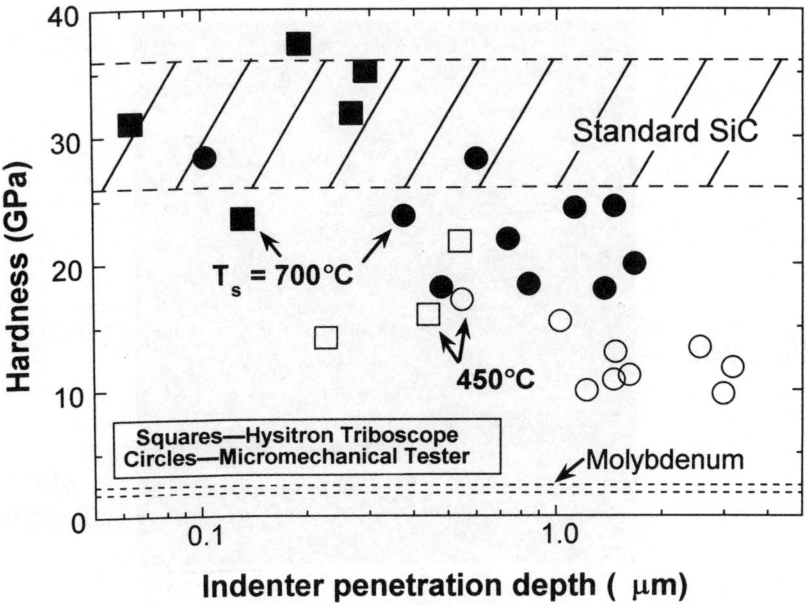

Fig. 8. Measured hardness versus nanoindenter penetration depth of two SiC films deposited at different substrate temperatures.

6.3.3. Layered titanium/titanium carbide film deposition

A coating consisting of alternating metallic-ceramic layers may have significantly enhanced fracture toughness, much like a cermet, compared with a pure ceramic coating. To demonstrate the feasibility of depositing layered nanocrystalline composites, alternating layers of titanium and titanium carbide were deposited by injecting titanium tetrachloride as the titanium souce and switching the carbon feedstock (methane) on and off. Fig. 9 shows an SEM micrograph of a cross-section of a film deposited using this approach. The film consists of five layers, each 3–4 μm thick, alternating between the carbide and the metal. Unfortunately analyses by XRD and by X-ray photoelectron spectroscopy suggest that the titanium is significantly oxidized, so that the goal of a metal-ceramic composite may not have been achieved. Modifications are currently underway to pretreat the plasma gases with a copper-based oxygen gettering system, in order to reduce the probability of oxidizing the titanium particles. In any case, the feasibility of depositing layered nanocomposites by HPPD has clearly been demonstrated.

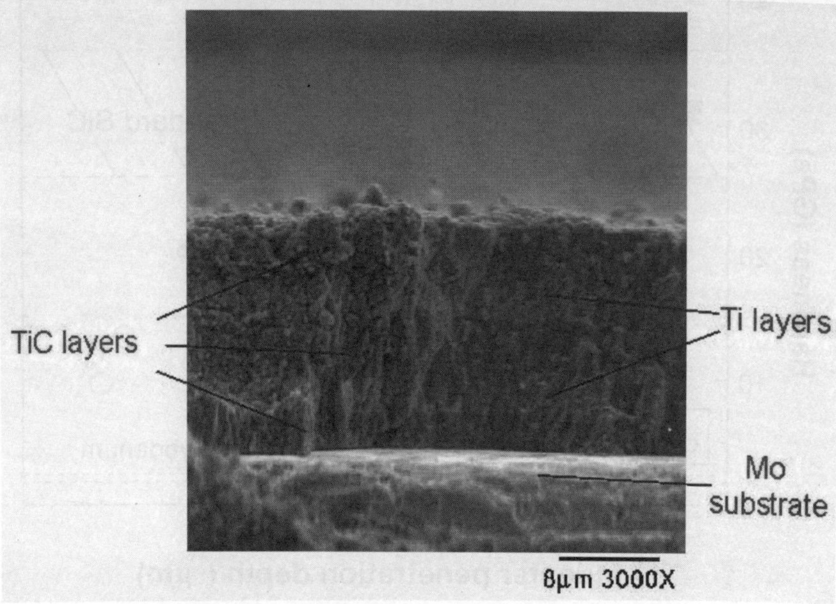

Fig. 9. Scanning electron micrograph of layered titanium–titanium carbide film.

7. Film Post-Treatment

The measured density of the as-deposited silicon carbide films ranged from 65–95% of the theoretical value of bulk SiC. Film post-treatment at elevated temperature and/or pressure is of interest for densification as well as for driving off adsorbed contaminants. Preliminary results have been obtained from several experiments in which nanocrystalline silicon carbide films deposited by HPPD were post-treated either by hot isostatic pressing (HIPing) or by vacuum sintering.

The HIPing experiments were carried out in a Patterson Instruments rock deformation instrument. Fig. 10 shows SEM micrographs (at different magnifications) of two post-treated silicon carbide films. The film on the left was HIPed at a peak pressure of 80 MPa, with an additional axial load of 25.5 MPa and a peak temperature of 900°C. Total HIP time was 105 min, with 10 min under peak conditions. The film on the right was HIPed at a peak pressure of 100 MPa, without additional axial load, and a peak temperature of 995°C. Total HIP time was 110 min, with 10 min under peak conditions. Comparing these two figures with micrographs of the same films before post-treatment, it is evident that the sample on the left experienced no grain growth (in fact, not even interparticle necking), whereas the sample on the right experienced considerable grain growth.

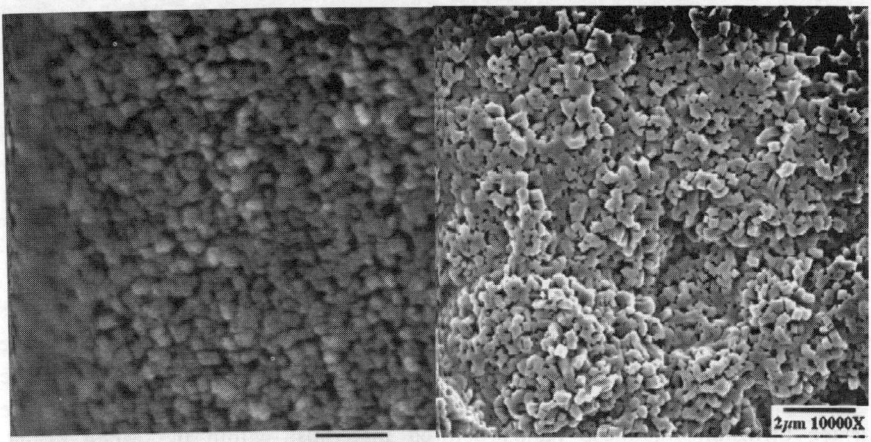

Fig. 10. Scanning electron micrographs of silicon carbide films post-treated by hot isostatic pressing. Left: HIPed at 80 MPa, with an additional axial load of 25.5 MPa and a peak temperature of 900°C. Right: HIPed at 100 MPa, without additional axial load, and a peak temperature of 995°C. Note the different magnifications of the two micrographs.

To clarify these results, two HPPD-deposited silicon carbide films were post-treated in a vacuum sintering oven. Both samples were treated for 10 min, one at 900°C, the other at 1000°C. SEM micrographs (at slightly different magnifications) of these vacuum-sintered films are shown in Fig. 11. The sample treated at 900°C shows little grain growth. The fact that some interparticle necking is evident, whereas none appeared in the sample HIPed at 900°C (Fig. 10), is possibly related to uncertainties in temperature measurement. In any case, the sample that was vacuum-sintered at 1000°C clearly shows considerable grain growth.

Fig. 11. Scanning electron micrographs of silicon carbide films post-treated by vacuum sintering. Left: sintered at 900 °C. Right: sintered at 1000°C. Note the different magnifications of the two micrographs.

Taking the above results together, it appears that grain growth for these films was activated at a temperature between 900 and 1000°C, and was unaffected by the applied pressures. Grain growth as a result of heating is caused by surface diffusion. As is clearly shown by these results, pressure alone does not cause surface diffusion for these nanostructured materials, within the range tested. However pressure can cause densification by plastic deformation. These preliminary results are thus consistent with previous observations [9] that nanostructured materials can potentially be sintered to full density without grain growth.

8. Focused Nanoparticle Beam Deposition of Patterned Microstructures

We recently [41] developed a modification of the HPPD process, in which the substrate is replaced by a set of "aerodynamic lenses" [42-44] which collimate the nanoparticles to form a beam. A substrate is placed downstream of the lenses. If the substrate is stationary, a high-aspect-ratio "tower" is deposited. Lines and two-dimensional patterns can be deposited by translating the substrate. This process could potentially be used for microfabrication of structures in microelectromechanical systems. The inherent nanocrystalline nature of these microstructures could confer important advantages with regard to mechanical, optoelectronic, or other properties.

The concept of generating narrow beams of nanoparticles to fabricate microstructures has been pursued by researchers in Japan [45-49], who developed a method termed "gas jet deposition," in which particle beams are produced by passing an ultrafine aerosol through extremely fine micronozzles, a typical nozzle diameter being 100 μm. Aerodynamic focusing, in contrast, allows one to generate particle beams with dimensions much smaller than the dimensions of the nozzles or orifices used, so that micronozzles are not required. For example, in our experiments the minimum flow dimension (the inner diameter of the exit nozzle of the aerodynamic lens system) is 1.85 mm, but the deposited line width is approximately 50 μm.

An "aerodynamic lens" is simply a plate with an orifice in the centre. The lens assembly used in these experiments consisted of five lenses in series, each with an orifice diameter of 2.26 mm. The lenses were spaced 47 mm apart, and the lens thickness was 0.3 mm. The pressure is about 333 Pa at the inlet of the lens assembly and 1.0 Pa in the chamber downstream of the lens assembly. The flow is therefore choked at the exit nozzle, whose diameter thus determines the flow rate through the lens assembly, and particles again deposit by inertial impaction.

Fig. 12 shows the predictions of a numerical simulation for both the gas streamlines and the trajectories of 20-nm-diameter (assumed spherical) silicon carbide particles for a lens assembly with the same geometry and operating conditions as in our experiments. For these simulations steady, laminar, compressible flow was assumed, and particle trajectories accounting for viscous drag but not Brownian diffusion were calculated. Diffusion would be expected to broaden the beam width, especially for particles smaller than about 10 nm in diameter.

As can be seen in Fig. 12, in passing through each orifice the gas contracts and then re-expands. Sufficiently small particles are able to follow the gas streamlines. Large particles accelerate radially towards the axis as the flow approaches the lens, and their high inertia projects them across the centreline to the opposite wall where they are lost. Particles of an intermediate size which, for our geometry and conditions includes 20-nm particles, are similarly accelerated towards the centreline as the flow approaches the orifice but, due to their smaller inertia they terminate their radial motion on a flow streamline closer to the axis than that on which they originated. Fig. 12 shows that the 20-nm particles entering the lens along various streamlines are progressively moved toward the centreline as each lens is passed, producing a tightly collimated beam at the nozzle exit. By careful design of the lens assembly it is possible to collimate particles within a specified range of sizes.

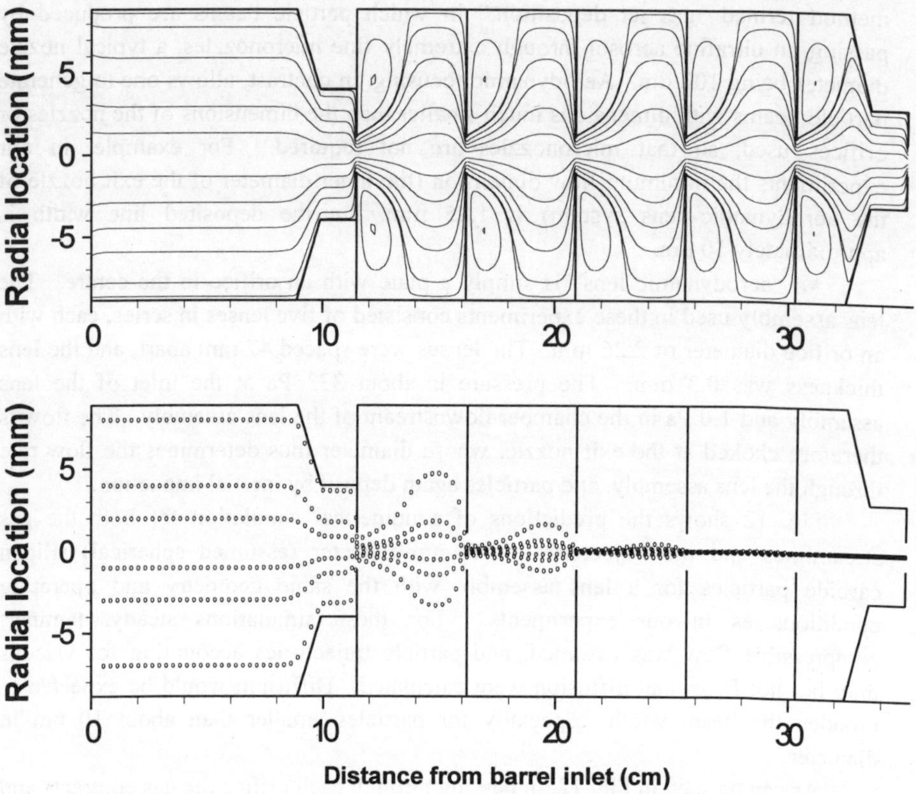

Fig 12. Gas streamlines (top) and trajectories of 20-nm silicon carbide particles (bottom), predicted by numerical simulation, for aerodynamic lens assembly with the same geometry and conditions used in experiments.

The flow in the chamber upstream of the aerodynamic lens assembly is expected to have relaxed to close-to-room temperature at the assembly inlet, which is located 75 cm downstream of the plasma expansion nozzle, and the deposition substrates were also at room temperature. Several substrate materials were tested, including aluminum, stainless steel, brass and glass. Adherent deposits formed on all of the metal substrates but not on glass.

With stationary substrates the particle beams deposit needle-shaped "towers," as seen for example in Fig. 13. Both silicon carbide and titanium towers were deposited. Linear growth rates of these towers were 1–2 μm/s. High resolution SEM micrographs of cross sections of the titanium towers show grain sizes of about 20 nm, which is similar to the results for the continuous film deposition.

In preliminary experiments with substrate translation we have demonstrated the feasibility of depositing lines and two-dimensional patterns. Fig. 14 shows a zig-zag pattern deposited by manual substrate translation. The minimum line width is approximately 50 μm. It can be seen that towers began to grow at a few locations where the substrate translation momentarily paused. With computer-controlled substrate rastering this process might be used for microfabrication of three-dimensional structures.

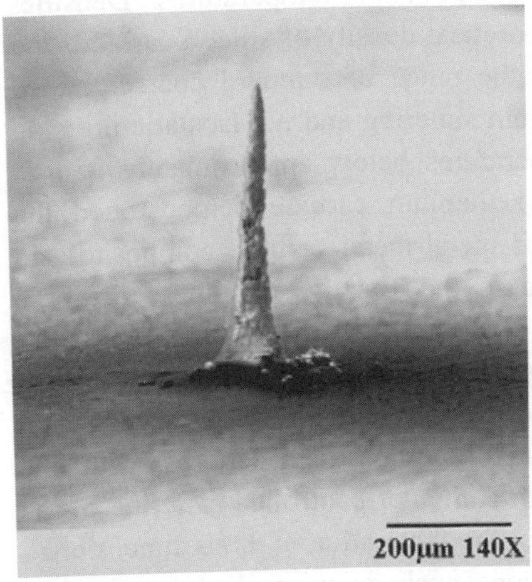

Fig 13. Silicon carbide "tower" deposited by focused particle beam.

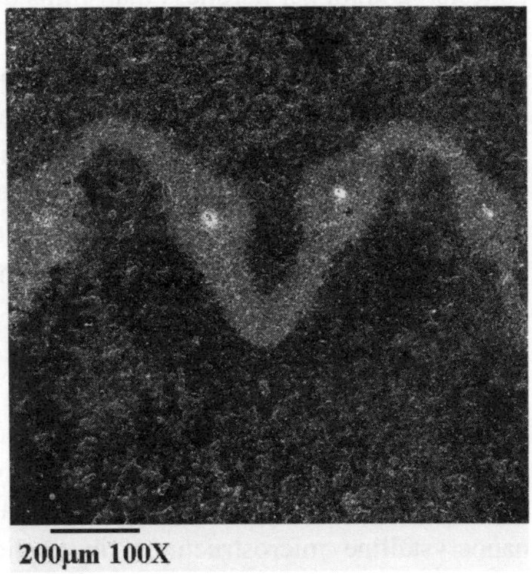

Fig.14. Pattern formed by focused beam of silicon carbide particles with manual translation of substrate [41].

9. Summary and Conclusions

Hypersonic plasma particle deposition is a new technology that offers a means of depositing nanocrystalline films at rates comparable to conventional thermal spraying and a few orders of magnitude higher than in chemical vapour deposition. The key elements of HPPD are that vapour-phase precursors are injected into a thermal plasma source, providing a high-rate source of nanoparticles; the plasma is expanded through a nozzle, providing a well-characterized, close to one-dimensional environment for particle nucleation and growth, and the flow is operated in the hypersonic region, so that particles as small as a few nm in diameter deposit by inertial impaction.

Results for silicon carbide deposition showed that several film properties were affected by substrate temperature. Films deposited at higher substrate temperatures (around 650–700°C) showed more crystallinity and had higher density, better adhesion and higher hardness than films deposited at lower temperatures. Densities of as-deposited films reached 95% of the theoretical density of silicon carbide, and measured hardness reached 37 GPa, above the range of standard coarse-grained SiC. Post-processing of these films by vacuum sintering and hot isostatic pressing indicated negligible grain growth for temperatures below approximately 950°C. Preliminary experiments with titanium and titanium carbide demonstrated the feasibility of using HPPD to deposit layered metal-metal carbide composites, an approach which has the potential to increase the fracture toughness of these nanostructured films.

In a modification of HPPD, the deposition substrate is replaced by an aerodynamic lens system, which focuses the nanoparticles to a collimated beam. This beam can be used to deposit lines and two-dimensional patterns, with line widths on the order of 50 μm. Together with computer-controlled rastering of the substrate, this new technology has potential for the fabrication of three-dimensional, nanocrystalline microstructures, in applications such as microelectromechanical systems.

Acknowledgments

The work reviewed in this paper covers the period 1993-2000. In addition to the authors of this paper we wish to acknowledge the contributions of former colleagues at the University of Minnesota, including M. Bench, D. Berns, J. Blum, C.-P. Chiu,

C. Fandrey, S. Fearon, D. J. Hansen, M. Hatch, S. Jones, M. Kelkar, H. J. Lee, B. Micheel, A. Neuman, L. Stachowicz, Q. Wei, and Z. Wong.

This work was partially supported by the U.S. National Science Foundation (current grant DMI-9871863), by the Center on Plasma-Aided Manufacturing and by the University of Minnesota Supercomputing Institute.

References

1. Siegel R. W. *et al*, *J. Mater. Res.* **3** (1988), 1367-1372.
2. Gleiter H., *Prog. Mater. Sci.* **33** (1989), 223-315.
3. Averback R. S. *et al*, *Mat. Res. Soc. Symp. Proc.* **153** (1989), 3-12.
4. Mayo M. J., Siegel R. W., Narayanasamy A. and Nix W. D., *J. Mater. Res.* **5** (1990), 1073-1082.
5. Siegel R. W., *Annu. Rev. Mater. Sci.* **21** (1991), 559-578.
6. Nieman G. W., Weertman J. R. and Siegel R. W., *J. Mater. Res.* **6** (1991), 1012-1027.
7. Gleiter H., in *Mechanical Properties and Deformation Behavior of Materials Having Ultra-fine Microstructures*, ed. Nastasi M., Parkin D. M. and Gleiter H. (Kluwer Academic Publishers, Dordrecht, 1993), 3-35.
8. Siegel R. W., *Mater. Sci. Eng.* **A168** (1993), 189-197.
9. Siegel R. W., in *Mechanical Properties and Deformation Behavior of Materials Having Ultra-fine Microstructures*, ed. Nastasi M., Parkin D. M. and Gleiter H. (Kluwer Academic Publishers, Dordrecht, 1993), 509-538.
10. Veprek S., *J. Vac. Sci. Technol. A* **17** (1999), 2401-2420.
11. Kong H. and Ashby M. F., *Acta Metall. Mater.* **40** (1992), 2907–2920.
12. Lau M. L., Jian H. G., Nüchter W. and Lavernia E. J., *Phys. Stat. Sol. A* **166** (1998), 257-268.
13. Wang Y. *et al*, *Wear* **237** (2000), 176-185.
14. Karthikeyan J. *et al*, *J. Am Ceram. Soc.* **81** (1998), 121-128.
15. Goswami R., Sampath S., Herman H. and Parise J. B., *J. Mater. Res.* **14** (1999), 3489-3492.
16. Rao N. P., Girshick S. L., McMurry P. H. and Heberlein J. V. R., U.S. patent no. 5,874,134 (1999).
17. Girshick S. L., *Plasma Sources Sci. Technol.* **3** (1994), 388-394.
18. Girshick S. L. *et al*, *J. Aerosol Sci.* **24** (1993), 367-382.
19. Rao N. *et al*, *J. Mater. Res.* **10** (1995), 2073-2084.

20. Rao N. *et al*, *Plasma Chem. Plasma Process.* **15** (1995), 581-607.
21. Hansen D. J., M. S. thesis, University of Minnesota, 1995.
22. Wang S. C. and Flagan R. C., *Aerosol Sci. Technol.* **13** (1990), 230-240.
23. Fernandez de la Mora J., Hering S. V., Rao N. and McMurry P. H., *J. Aerosol Sci.* **21** (1990), 169-187.
24. Liu B. Y. H., Whitby K. T. and Pui D. Y. H., *J. Air Poll. Control Assoc.* **24** (1974), 1067-1072.
25. Stolzenburg M. and McMurry P. H., *Aerosol Sci. Technol.* **14** (1991), 48-65.
26. Micheel B. *et al*, in *Proc. 12th Intl. Symp. Plasma Chem., Minneapolis, Aug. 21-25, 1995*, v. 3, 1379-1384.
27. Seinfeld J. H. and Pandis S. N., *Atmospheric Chemistry and Physics: From Air Pollution to Climate Change* (John Wiley & Sons, New York, 1998).
28. Fernandez de la Mora J., *J. Chem. Phys.* **82** (1985), 3453-3464.
29. Fernandez de la Mora J. and Schmidt-Ott A., *J. Aerosol Sci.* **24** (1993), 409-415.
30. Rao N. P. *et al*, *J. Aerosol Sci.* **29** (1998), 707-720.
31. Lai F. S., Friedlander S. K., Pich J. and Hidy G. M., *J. Coll. Interface Sci.* **39** (1972), 395-405.
32. Rao N. P. *et al*, *Nanostruct. Mater.* **9** (1997), 129-132.
33. Blum J. *et al*, *J. Nanoparticle Research* **1** (1999), 31-42.
34. Bieberich M. T. and Girshick S. L., *Plasma Chem. Plasma Process.* **16** (1996), 157S-168S.
35. *CFD-ACE Command Language Reference Manual*, Version 2.1, CFDRC Report GR-97-30, CFD Research Corp., Huntsville, AL, 1997.
36. Kee R. J., Report SAND87-8215B, Sandia National Laboratories, 1986.
37. Neuman A. *et al*, *IEEE Trans. Plasma Sci.* **27** (1999), 46-47.
38. Oliver W. C. and Pharr G. M., *J. Mater. Res.* **7** (1992), 1564.
39. Askeland D. R., *The Science and Engineering of Materials* (PWS-Kent Publishing Co., Boston, 1989).
40. Barsoum M., *Fundamentals of Ceramics* (McGraw-Hill, New York, 1997).
41. Di Fonzo F. *et al*, *Appl. Phys. Lett.* **77** (2000), in press.
42. Liu P. *et al*, U.S. patent no. 5,270,542 (1993).
43. Liu P., Ziemann P. J., Kittelson D. B. and McMurry P. H., *Aerosol Sci. Technol.* **22** (1995), 293-313.
44. Liu P., Ziemann P. J., Kittelson D. B. and McMurry P. H., *Aerosol Sci. Technol.* **22** (1995), 314-324.

45. Kashu S., Fuchita E., Manabe T. and Hayashi C., *Jpn. J. Appl. Phys.* **23**, Part 2 (1984), L910-L912.
46. Oda M. *et al*, *MRS Symp. Proc.* **286** (1993), 121-130.
47. Schroth A., Maeda R., Akedo J. and Ichiki M., *Jpn. J. Appl. Phys.* **37** (1998), 5342-5344.
48. Akedo J., Ichiki M., Kikuchi K. and Maeda R., *Sensors Actuators A* **69** (1998), 106-112.
49. Akedo J. and Lebedev M., *Jpn. J. Appl. Phys.* **38** (1999), 5397-5401.

CHAPTER SEVEN

AEROSOL-BASED FLAME SYNTHESIS: A MICROREACTOR FOR SILICA NANOPARTICLES

Karsten Wegner and Sotiris E. Pratsinis

Institute of Process Engineering, ETH Zürich
CH-8092 Zürich, SWITZERLAND

1. Introduction

Flame aerosol reactors are routinely used for the inexpensive production of submicron-sized commodities such as carbon blacks, pigmentary titania and fumed silica and have high potential for manufacture of other nanoparticles (Pratsinis, 1998). The annual production volume of the flame industry is several million metric tonnes and aerosol reactors produce nanoparticles at rates in the order of 100 metric tonnes per day and higher. Compared with wet chemical production methods, these gas-phase processes are advantageous since they do not involve the tedious and expensive steps of solid-liquid separation, washing and drying, and they avoid the use of high liquid volumes and surfactants (Pratsinis and Mastrangelo, 1989). As a consequence, the production of pigmentary titania by oxidation of $TiCl_4$ in a flame (chloride process) is replacing the wet-chemistry-based sulphate process in many areas of the titania industry (Ulrich, 1984). The manufacture of uranium dioxide by oxidation of uranium hexafluoride is another example of a flame process with good potential to replace its wet precipitation counterpart (Ulrich, 1984).

Powders produced by gas-phase combustion synthesis are usually aggregated and may be described as 'fractal-like'. These aggregates are polydisperse and comparably large in size with dimensions of up to two orders of magnitude higher than the primary particles (Wooldridge, 1998). Processes for the synthesis of non-aggregated nanoparticles therefore still operate mainly in the liquid phase, as with the synthesis of monodisperse transition metal oxide colloids by sol-gel methods (Livage et al., 1988). Consequently, the aim in aerosol flame reactor research is to develop flexible processes which allow the synthesis of a broad variety of nanoparticles ranging from non-aggregated spheres to fractal-like aggregates through control of the operating conditions. The size and morphology of flame-generated particles depend on the structure and properties of the flame. The most important parameters are the flame temperature field and the particle residence time. In diffusion flame reactors for example, these can be controlled through the mixing of precursor and oxidant, as was shown by the CFD-calculations of Johannessen et al. (2000). Recent studies have shown that by changing these parameters the properties of nanoparticles can be well controlled in laboratory-scale diffusion flame aerosol reactors at production rates from a few milligrams up to 1 kg/h.

Compared with premixed flames, diffusion flame reactors offer the advantage of high flexibility in product qualities by controlling the reactant gas composition over broad ranges. Zhu and Pratsinis (1996) investigated the mode of reactant gas mixing in the synthesis of titania powders with a diffusion flame reactor. They observed that oxygen stream composition drastically changed the product morphology from highly ramified aggregates to perfectly spherical particles. Fractal-like particles with primary particle sizes down to 11 nm, and non-aggregate, spherical nanoparticles with diameters of 100 to 200 nm, could be produced. Zhu and Pratsinis (1997) succeeded also in synthesizing spherical silica particles in the same size range which exhibited only a low degree of aggregation. Recently, Glumac et al. (1998) reported the production of non-aggregated silica, titania and alumina in a low pressure flame at rates up to 50 g/h. Using a double diffusion flame, Kammler and Pratsinis (1999) succeeded in producing silica nanoparticles with well defined properties at rates of up to 130 g/h. Zachariah and Semerjian (1990) investigated the influence of the silica precursor material on particle growth in a counterflow diffusion flame. They found that particle formation from silane occurs earlier and at lower temperatures than from hexamethyldisiloxane (HMDSO) and tetramethyl-silane (TMS). Briesen et al (1998a) who investigated silica formation in a diffusion flame from $SiCl_4$, HMDSO and octamethyl-cyclotetrasiloxane (OMCTS), found that the specific surface area of the nanoparticles could be controlled in a range of 30 to 250 m^2/g depending on the choice of precursor as well as the oxidant composition and flow rate. More accurately, they found a linear relationship between the specific surface area of the product powder and the adiabatic temperature of a premixed flame.

The introduction of additives or dopants into particle-generating flames is a further means to control the morphology and the crystallinity of product nanoparticles. Vemury and Pratsinis (1995a) added $SiCl_4$, $SnCl_4$, and $AlCl_3$ to a titanium tetrachloride precursor stream and found that the introduction of the silica precursor inhibits the titania phase transformation from anatase to rutile, decreases the primary particle size and, as a result, increases the specific surface area. However, when $SnCl_4$ or $AlCl_3$ were used as dopants (Vemury and Pratsinis, 1995a), the phase transformation of anatase to rutile was enhanced and the specific surface area was decreased. Fotou et al. (1995) investigated the role of ferrocene in the flame synthesis of silica and found that the presence of this additive increased the specific surface area of silica up to 150% and removed the coarse tail of the silica aggregate size distribution. Retardation of coagulation by charging was a possible explanation for these results.

The generation of charges by external electric fields offers another tool for control of characteristics of flame-made powders. Vemury and Pratsinis (1995b) generated titania nanoparticles by oxidation of $TiCl_4$ in a methane/air double diffusion flame in the presence of an electric field created by two needle electrodes across the flame. They showed that the specific surface area of the product particles decreased proportionally to the applied electric field strength and could be finely controlled. The influence of an electric field on the synthesis of silica particles in a diffusion flame reactor was investigated by Briesen et al. (1998b) and Kammler and Pratsinis (2000) for production rates up to 30 g/h and 87 g/h respectively. Both studies found that the specific surface area of the composite silica-carbon product particles, and also the powder composition, could be precisely controlled.

The studies mentioned above have shown that laboratory scale diffusion flame aerosol reactors with diameters of a few centimeters are well suited to control the size and morphology of the product particles. Little is known however about the influence of the size of co-flow diffusion burners on the properties of the generated powders. Wießmeier and Hönike (1996) showed that the performance of chemical reactors can improve if the dimensions of the reactor are reduced to microscale. Some microfabricated compounds that perform the same standard unit operations as those present in large chemical processing plants have already been developed including pumps, heat exchangers, separation units, and chemical reactors (Wegeng and Dorst, 1998). This new class of chemical process equipment has significantly different features from classical chemical processing units because microscale phenomena like heat and mass transport are exploited more effectively (Wegeng et al., 1996). Chemical microreactors in particular exhibit improved mixing of flows, extremely rapid heat and mass transfer, and higher processing rates (Burns et al., 1997) and as a result process variables can be controlled more precisely than in large scale units leading to higher yields and better product quality. Another advantage of microchemical systems is the high production flexibility for small amounts of chemicals on demand and on-site (Ehrfeld et al., 1997). A scale-up can be done easily by operating microreactor units in parallel and these systems are therefore especially suited for distributed processing applications (Wegeng and Dorst, 1998).

Until now, the concept of microreactors has not been extended to aerosol reactors despite the potential improvement in reactant mixing and transport properties in diffusion flames. The characteristics of isothermal double coaxial jets on which most diffusion flame burners are based may illustrate the potential improvement of reactant mixing in aerosol microreactors. A schematic of the near nozzle flow field of a double concentric jet is shown in Figure 1, which refers to a typical condition with higher outer jet velocity. Due to the entrainment requirements of free jets regions of pure nozzle fluids A and B, as well as areas of mixed gases, form in the initial merging zone downstream of the nozzle. The conical region in front of the centre tube containing fluid from the centre nozzle only is called the primary or inner potential core (Ko and Kwan, 1976). The corresponding region in front of the outer tube containing just fluid from the annular jet is named the secondary or outer potential core. The initial merging zone ends roughly at the tip of the outer potential core. Between the two potential cores lies the inner mixing region containing fluids A and B. In a typical diffusion flame with fuel provided through the centre and oxidant through the annular jet, the flame front would be established within this region at locations where the mixture of fuel and oxidant is stoichiometric. The secondary mixing zone starts at the outer boundary of the annular jet and contains annular jet fluid as well as gas entrained from the surrounding environment. Downstream of the initial merging zone is the intermediate zone, where mixing of the flows from the two upstream mixing regions occurs. When the two jets have combined in the fully merged zone, they behave essentially as a single jet.

The length of the inner potential core may be taken as an indicator for the effectiveness of mixing in a double coaxial jet. Mixing between the central and annular streams and therefore the length of the inner potential core is influenced by the outer and inner exit velocities and diameters, the thickness of the inner duct wall and the outflow conditions of the jets (Talamelli and Petagna, 1994). Shorter distances between the two jet centrelines, and thus faster mixing, can be realized by reducing the tube diameters. Villermaux et al. (1994) reported a decrease of the inner potential core length proportional to the inner nozzle diameter for a given velocity ratio r between the streams of a fully turbulent coaxial jet. Increasing the velocity ratio also decreased the core length proportionally. Chigier and Béer (1964) investigated the influence of the velocity ratio r on gas mixing for an isothermal double concentric jet. The concentration of central nozzle fluid at the centre axis of the jet decreased to 10% of its initial value after a distance of 16 nozzle diameters for a velocity ratio of $r = 3.3$, but it reached this value after only 4 nozzle diameters for r = 9.2, demonstrating the importance of the velocity ratio for reactant mixing.

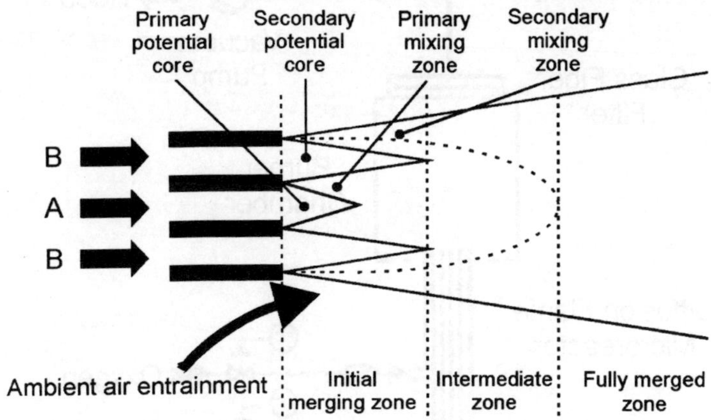

Figure 1. Schematic of the flow field in a double concentric jet. The mean exit velocity of the outer annular jet is higher than that of the central jet.

In this work these concepts are applied to a diffusion flame microreactor with small nozzle diameters and high outer-to-inner jet velocity ratios. The microreactor is performance-tested by the example of the production of silica nanoparticles from hexamethyldisiloxane as precursor and checked for its ability to produce nanoparticles over a broad range of sizes and structures.

2. Experimental

A schematic of the experimental set-up is shown in Figure 2. The diffusion flame microreactor (Figure 3) consisted of five concentric stainless steel tubes, the outer tube having an inner diameter of 8 mm while the i.d. of the inner tube was 1.8 mm. The gap between the tubes was about 0.8 mm. An argon stream saturated with the precursor hexamethyldisiloxane (HMDSO, Fluka, purity >99%) was introduced into the methane/oxygen/nitrogen diffusion flame of the microreactor. High purity methane, oxygen, nitrogen and argon (Pan Gas, purity >99.999 %) was delivered from cylinders with the flow rates being monitored by calibrated mass flow controllers (Bronkhorst EL-Flow F201).

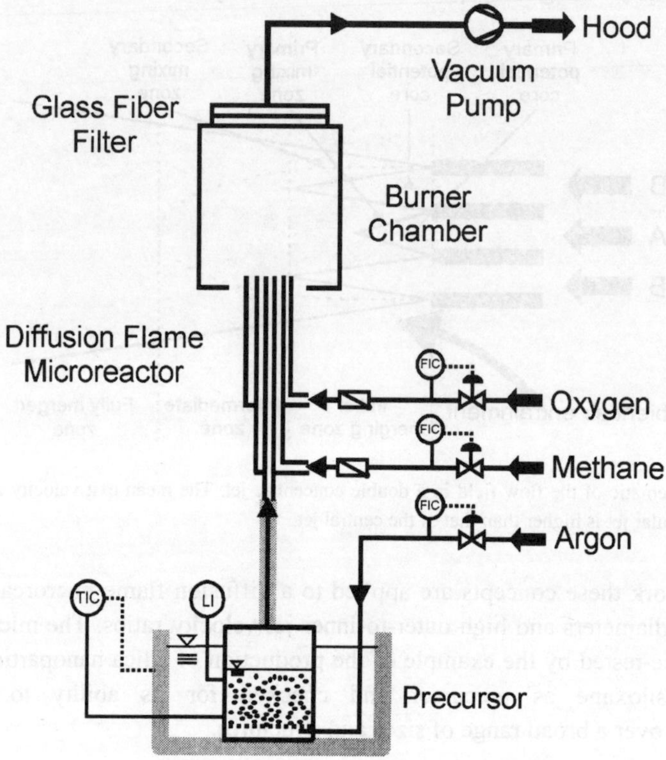

Figure 2. Schematic of the experimental set-up with the diffusion flame microreactor operated in the C-flame configuration.

Before entering the reactor, argon was bubbled through the precursor flask placed in a thermostat-controlled bath (Huber unistat cc) and was saturated with the organometallic compound. By setting the bath temperature, controlled amounts of precursor could be delivered into the reactor at an argon flow-rate of 300 cm^3/min. At 25°C the HMDSO delivery rate was 0.73×10^{-3} mol/min if 100 % saturation of the argon stream was achieved. This was confirmed by weighing the flask before and after an experiment. The precursor delivery tube and the burner were heated with heating tapes 20°C above the bath temperature to avoid condensation of precursor. The precursor laden argon stream was delivered through the centre tube, methane through the first annulus and the oxidant through the second annulus. This flame configuration corresponds to a C-flame (Pratsinis et al., 1996).

Figure 3. Diffusion flame microreactor and a Swiss five Rappen coin of 1.7 cm diameter. The outer two tubes of the microreactor are not used for powder production.

In another flame configuration the methane joined the precursor stream in the inner tube. The oxidant was delivered through the third tube while nitrogen flowed through the second tube in order to lift the flame a few millimeters from the burner mouth, thus preventing the deposition of particles. This flame configuration is hereafter referred to as the D-flame. In all experiments the methane flow rate was kept constant at 500 cm^3/min. The flow rate of the oxidant stream was varied between 1000 and 12000 cm^3/min. These flow rates gave a range of 1.6 to 19.2 for the velocity ratio r between the outer annular and inner center jets. Either pure oxygen or an oxygen/nitrogen mixture with the molar ratio 2:1 were used as oxidant. Experiments with higher nitrogen fractions were not carried out as the flame could not be stabilized and was in fact extinguished at oxidant flow rates larger than 6000 cm^3/min.

The flame was surrounded by a quartz chimney (glass cylinder, i.d. = 140 mm) in order to achieve stable burning. A stainless steel filter holder with a glassfiber filter (Whatman GF/A) was mounted on top of this cylinder. The product particles were collected on the filter by the aid of a vacuum pump (Vacuubrand RE5). Each experiment was reproduced twice.

The specific surface area (SSA) of the collected powders was analyzed twice for each sample by nitrogen adsorption at 77K employing the BET isotherm (Micromeritics GEMINI 2360). The results were cross-checked by recording a full adsorption isotherm (Micromeritics ASAP 2010 Multigas system). Samples of the synthesized powders for analysis by transmission electron microscopy (TEM, Hitachi H600, 100 kV) were taken in the gas phase with the aid of a thermophoretic sampler (Dobbins and Megaridis, 1987) and a carbon-coated copper grid was inserted for 200 ms into the aerosol stream at the location of the filter on the center axis of the burner. Thermophoretic sampling was also carried out in-situ: the C-flame operated with an oxygen flow rate of 2000 cm^3/min. The sampling locations were at the centre axis of the flame at distances of 10, 30, and 50 mm from the burner mouth. The grid remained 50 ms in the flame.

Centreline temperatures of the flames were recorded with a 0.5 mm thin wire Pt-Rh thermocouple (Type B, Omega Engineering) in the absence of a precursor and were corrected for radiation loss. The thermocouple could be positioned precisely in the rotational symmetric flame by moving the burner with an x-y positioning table (Föhrenbach GmbH). Heat loss by radiation was estimated using an emissivity of 0.18 for a Pt wire (Lide, 1997) and a temperature of 100°C at the walls of the combustion chamber. Heat transfer by forced convection to the 1.4 mm bead of the thermocouple was calculated with the outlet velocity of the centre jet in lack of gas velocities in the flame. Those are assumed to be higher than the outlet velocity due to acceleration of the flow at the centreline by the high velocity annular jet and thermal expansion of the gas and therefore the reported data overestimates the flame temperature by 50 to 100 K. The combustion of HMDSO is an exothermic reaction with an enthalpy of combustion of -1409 kcal/mol (Arkles, 1995). For the investigated system, this accounts for about 20% of the methane reaction heat: thus temperatures of the precursor-laden flame are higher than in the methane flame.

Gas mixing in the triple coaxial jet of the microreactor was investigated by computational fluid dynamics, employing the commercial software Fluent 5.2. T; an isothermal (T = 300K) coaxial jet with the dimensions of the microreactor was simulated for atmospheric pressure, neglecting chemical reactions. The argon and methane flow rates were the same as in the powder production experiments of the C-flame. Oxygen flow rates of 2000, 4000 and 6000 cm^3/min were used. Due to the rotational symmetry of the coaxial jet, the simulation was carried out with a two-dimensional grid with a symmetry axis. A k-ε turbulence model was applied for the calculation. Coaxial jet mixing was also simulated for a larger reactor (outer diameter: 20 mm) used by Briesen et al. (1998a) for synthesis of silica nanoparticles. Conditions applied were the same as for the microreactor, with following gas flow rates: argon (center tube): 250 cm^3/min, methane (first annulus): 400 cm^3/min, oxygen (second annulus): 2500, 3800 and 5500 cm^3/min.

3. Results and Discussion

3.1. Jet mixing

Velocity vectors of the near nozzle region of the microreactor are shown in Figure 4. along with concentration profiles of argon for oxygen flow rates of 2000, 4000 and 6000 cm^3/min. For the lowest oxygen flow rate, the outlet velocity of the annular oxygen jet is 6 m/s and therefore about 3 times higher than the exit velocities of the methane and argon streams, which are 2.0 and 1.8 m/s, respectively. The annular jet entrains environmental air at its outer boundary and draws fluid from the two central jets at its inner boundary, radially deflecting the centre jets (Fig. 4a). The lack of fluid in the centre region results in a zone of subatmospheric pressure which deflects the annular jet toward the axis (Chigier and Béer, 1964). Increasing the oxygen flow rate by a factor of two (Fig. 4b) increases the entrainment requirements of the outer annular jet; within 0.5 mm from the nozzle exit, the methane and argon streams are deflected outward and drawn into the oxygen jet. The resulting lack of fluid in the central region forces the annular jet to draw fluid from itself to satisfy its entrainment requirements at the inside boundary. A toroidal recirculating vortex is set up with its centre at 1.5 mm from the nozzle exit plane. A zone of subatmospheric pressure associated with this vortex deflects the annular jet towards the axis. Increasing the oxygen flow rate to 6000 cm^3/min increases the length and strength of the recirculating vortex, as is indicated by steeper velocity gradients (Fig. 4c).

The location of the vortex center has moved to 1mm from the exit plane. All centre nozzle fluid is entrained by the annular oxygen jet after only 0.25 mm.

The influence of increasing oxygen flow rate on gas mixing is demonstrated by argon concentration profiles presented in Figure 4, showing lines of constant argon molar fraction. At an oxygen flow rate of 2000 cm^3/min an inner potential core of 1.5 mm length containing pure argon can develop (Figure 4a). The centreline molar fraction of argon is reduced to 0.5 after about 4 mm. The inner core does not form when the oxygen flow rate is increased to 4000 cm^3/min (Figure 4b). The center nozzle fluid is immediately drawn toward the annular oxygen stream as is indicated by the radial deflection of the lines of constant argon concentration. A centreline argon molar fraction of 0.5 is found at 0.8 mm from the nozzle exit, showing that the intensity of gas mixing increases with increasing outer- to-inner jet velocity ratio: further increase of the oxygen flow rate to 6000 cm^3/min reduces this distance to about 0.3mm. The steep concentration gradients indicate that fast mixing of the oxygen and argon streams takes place.

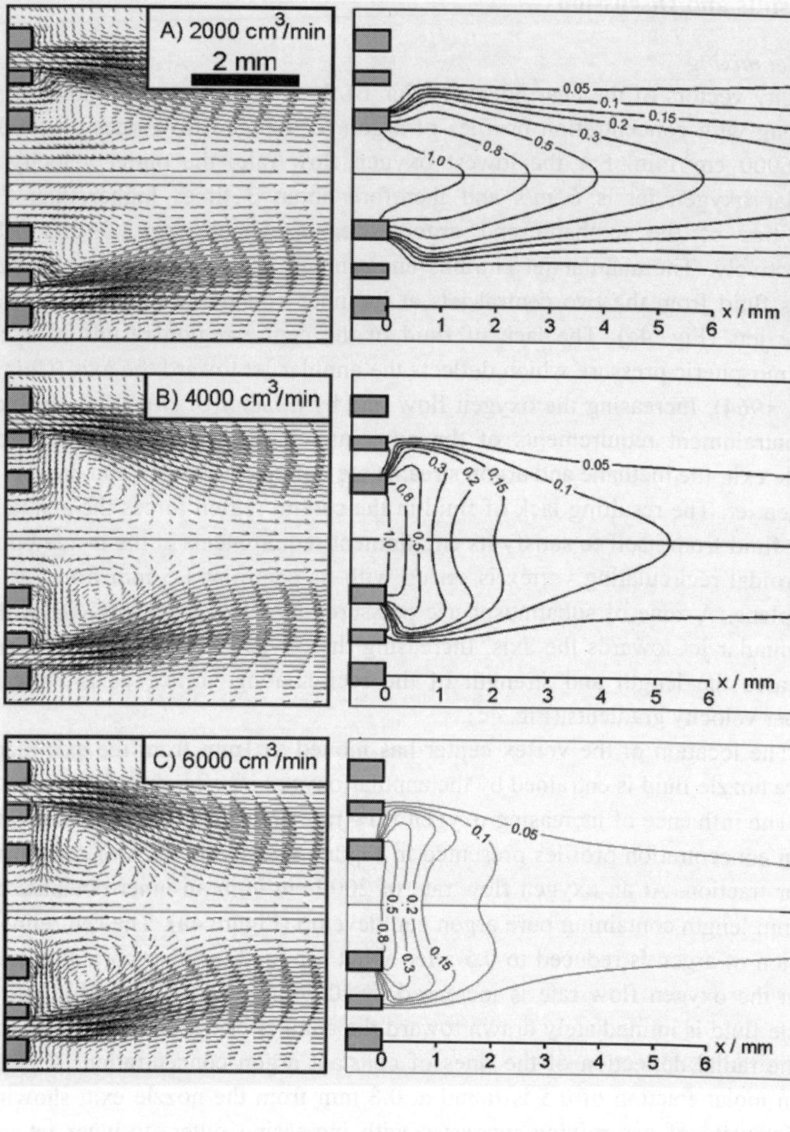

Figure 4. **Velocity vectors (left) of the near nozzle region of the present microreactor calculated by computational fluid dynamics along with lines of constant argon molar fraction (right). The oxygen flow rates through the outer annular jet are a) 2000, b) 4000 and c) 6000 cm³/min.**

Velocity vectors and argon concentration profiles of the reactor of Briesen et al. (1998a) are presented in Figure 5. At an oxygen flow rate of 2500 cm^3/min, giving an outlet velocity of 0.5 m/s and an outer-to-inner jet velocity ratio of about 2.5, the lack of entrainment fluid for the annular jet results in the backstreaming of centre nozzle fluid (Fig. 5a). A potential inner core of about 3 mm length can however still develop and the molar fraction of argon on the centreline is decreased to 0.5 after 10 mm, compared with 4 mm in the case of the microreactor. Increasing the outer annular jet flow rate to 3800 cm^3/min (Fig. 5b) leads to the formation of a toroidal recirculating vortex in the center region. Unlike the case of the microreactor where the vortex region consists mainly of premixed annular jet fluid, it consumes mainly argon in the jet configuration of Briesen et al. (1998a). As a result an inner potential core can develop and the centreline argon mole fraction of 0.5 is reached after 5 mm, compared with 0.8 mm for the microreactor, and thus mixing of argon and oxygen is assumed to be less intense. For the highest oxygen flow rate of Briesen at al. (5500 cm^3/min, Fig. 5c) a flow field similar to the one observed with the microreactor for oxygen flows of 4000 and 6000 cm^3/min is established, however because of the larger size of the reactor, argon concentration gradients are less steep and a centreline argon mole fraction of 0.5 is reached after about 2 mm.

3.2. Flame characterization

Figure 6 gives the centreline temperature profiles of the C-Flame with oxygen as oxidant (filled symbols) and oxygen/nitrogen ratios of 2:1 (open symbols). In both cases the maximum flame temperature was decreased when the flow rate of oxidant was increased. For pure oxygen, the maximum temperatures were 2400, 2030, and 1850 K for oxygen flow rates of 4000, 6000, and 8000 cm^3/min respectively. When oxidant streams of the same flow rates with O_2:N_2 = 2:1 were used, the maximum flame temperatures were approximately 2100, 1850 and 1800 K.

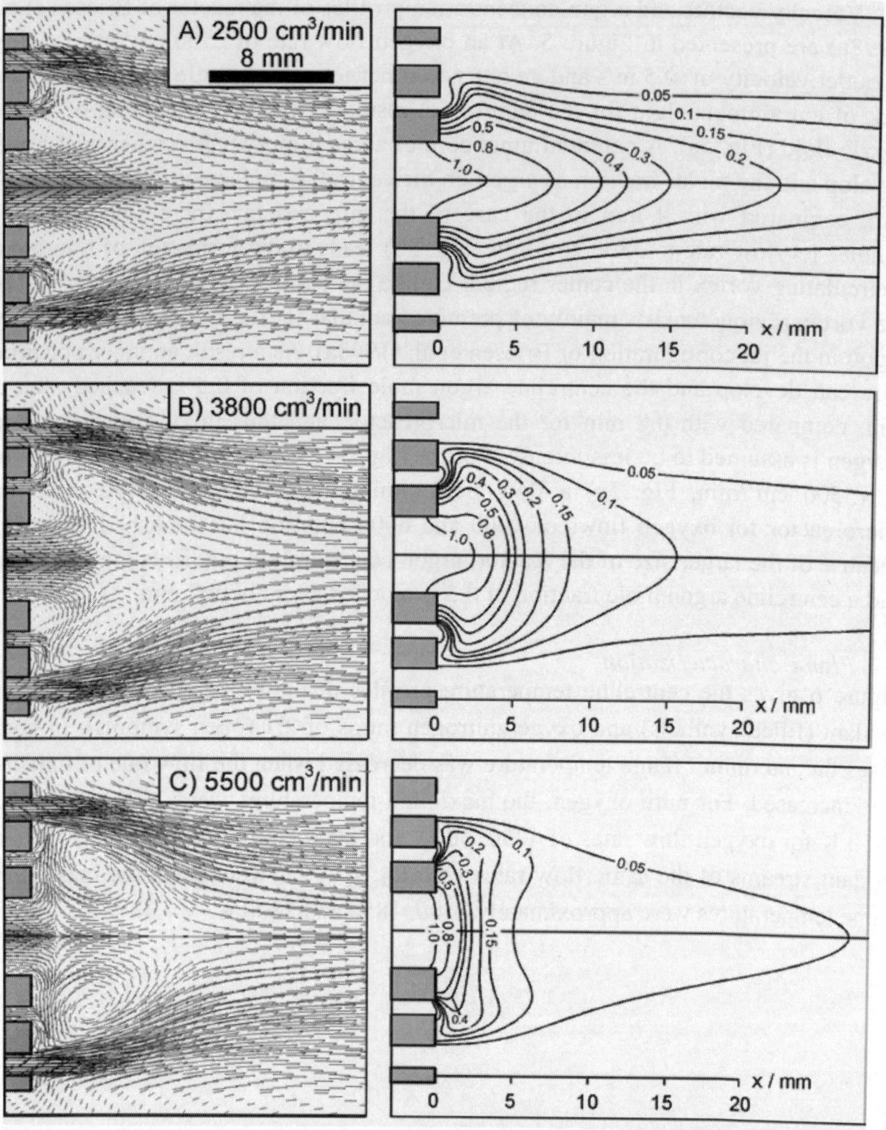

Figure 5. Velocity vectors and lines of constant argon molar fraction calculated by computational fluid dynamics for the reactor of Briesen et al. (1998a). Oxygen flow rates through the outer annular tube are a) 2500, b) 3800 and c) 5500 cm^3/min.

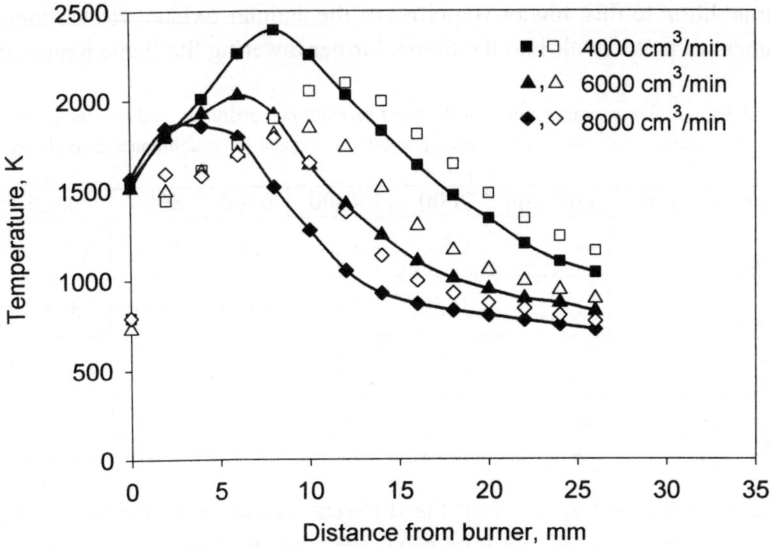

Figure 6. Temperature profiles of the C-flame measured at the flame centerline in absence of precursor for pure oxygen flow rates of 4000, 6000, and 8000 cm^3/min (filled symbols) and for an oxidant composition of O_2:N_2=2:1 (open symbols).

Temperatures of flames with oxidant flow rates of 2000 cm^3/min could not be recorded as the temperature exceeded the melting point of the thermocouple alloy. The location of the maximum flame temperature with respect to the burner mouth decreased with increasing oxidant flow rate. With pure oxygen as oxidant, that maximum was at 8, 6, and 4 mm while for O_2:N_2 = 2:1 the maximum was at 12, 10, and 8 mm for oxidant flow rates of 4000, 6000 and 8000 cm^3/min respectively. The addition of nitrogen to the oxygen stream dilutes the oxygen concentration in the flame and lowers the flame temperature up to 300 K, delaying the combustion of the fuel. In general, lower oxygen concentrations lead to longer and colder flames. This is explained by gas cooling and mixing.

The cooling effect is shown by comparing the adiabatic flame temperatures given in Table 1. With increasing excess air number λ, defined as the ratio of oxygen present in the flame over the amount of oxygen necessary for stoichiometric combustion, the adiabatic flame temperature decreases from 3650 K (2000 cm^3/min O_2) to 1611 K (8000 cm^3/min O_2) since unused oxygen dissipates the generated heat. In addition to this, higher velocities of the annular oxidant stream increase the entrainment of ambient air into the flame, further lowering the flame temperature.

Table 1. Adiabatic flame temperatures for different reactant compositions in the C-flame in absence of precursor. λ = (Amount of oxygen provided) / (Amount of oxygen for stoichiometric combustion).

Oxidant flow rate	cm^3/min	2000	4000	6000	8000	10000	12000
λ (pure O_2)	--	1.6	3.3	4.9	6.6	8.2	9.9
$T_{ad.}$ (pure O_2)	K	3650	2495	1939	1611	1395	1241
λ (O_2:N_2 = 2:1)	--	1.1	2.2	3.3	4.4	--	--
$T_{ad.}$ (O_2:N_2 = 2:1)	K	3720	2525	1962	1631	--	--

The mixing effect arises from the different velocities of the fuel and oxidant streams. The exchange of mass between two parallel fluid streams is known to increase with increasing velocity ratio of the streams (Günther, 1974). The velocity of the inner argon and methane streams at the outlet of the microreactor is 1.8 m/s. When the flow rate of the oxidant stream is increased from 2000 to 6000 cm^3/min, its outlet velocity rises from 6.0 to 18.1 m/s and the velocity ratio r, defined as the velocitiy of the annular jet to the centre jet increases from 3.2 to 9.6 respectively. As was shown by computational fluid dynamics, this results in faster mixing of the reactants for the concentric diffusion flame for higher oxidant flow rates, leading to a shift of the maximum flame temperature to shorter distances from the burner mouth.

Similar, but less pronounced effects of the oxidant flow rate on the maximum flame temperature were observed by Zhu and Pratsinis (1997) and Briesen et al. (1998a). They however first reported an increase in the maximum flame temperature when the oxidant flow rate was increased from 2500 to 3800 cm^3/min which is followed by lower temperatures when the flow rate is further enlarged. The initial increase in flame temperature was attributed to the fact that flames with oxygen flow rates smaller than 3800 cm^3/min were understoichiometric.

Unlike the situation in the microreactor, a significant influence of the oxidant flow rate on the location of the maximum flame temperature was not observed by Briesen et al. (1998a). While the temperature maximum of their flame stays at about the same position for oxygen flow rates of 2500 and 5500 cm^3/min, a noticeable shift is observed for similar flow rates in the microreactor: the different effectiveness of reactant mixing resulting from varying oxygen-to-fuel velocity ratios and different burner size accounts for this observation. At O_2 flow rates of 2000 and 6000 cm^3/min the oxygen to fuel velocity ratios are 3.2 and 9.7 for the microreactor and 2.5 and 5.5 for the reactor used by Briesen et al. (1998a). Intense reactant mixing observed for the microreactor within the first millimeters from the nozzle exit is not achieved with the larger reactor for the employed flow rates; thus a significant shift of the temperature maximum to short distances from the burner mouth cannot be achieved.

Figure 7 shows centreline temperature profiles of the D-flame configuration for different oxidant flow rates and compositions. The effect of the oxidant flow rate on the flame temperature is comparable to that of the C-flame but maximum temperatures are about 200 to 350 K lower than those of the corresponding C-flames. These results differ from those of Pratsinis et al. (1996), who recorded higher flame temperatures for the D-configuration than the C-configuration of their co-flow methane/air diffusion flame. The flowfield of the reactor of Pratsinis et al. (1996) is different from the one used in the microreactor here as they did not introduce nitrogen between the methane and oxygen streams. The additional nitrogen stream imposes a diffusion barrier for the species taking part in the combustio,n leading to longer and colder flames. This effect also explains the observation that the maximum flame temperatures of the D-flames is shifted to much longer distances from the burner mouth compared with the C-flames, as can be seen in Figures 6 and 7.

3.3. Nanoparticle synthesis

The influence of the oxidant flow rate on the specific surface area (SSA) of fumed silica nanoparticles synthesized with the diffusion flame microreactor operated in C-flame configuration, is shown in Figure 3.8 for the two oxidant compositions. The SSA of the silica particles increases from 40 to about 200 m^2/g when the oxygen flow rate is increased from 1000 to 12000 cm^3/min. The corresponding average BET-equivalent particle diameters range from 63 to about 13 nm, if spherical particles are assumed. The particle SSA follows a rather steep increase for oxygen flow rates up to 5000 cm^3/min and thereafter approaches an asymptotic value of 200 m^2/g at 8000 - 12000 cm^3/min. For flow rates up to 9000 cm^3/min the standard deviation of the measurements is smaller than 2% of the average specific surface area, indicating that the primary particle size can be precisely controlled with the microreactor. The shape of the curve SSA vs. oxidant flow rate is similar when one third of the oxygen is substituted by nitrogen but the maximum specific surface area of the silica powders is only 150 m^2/g for oxidant flow rates of 7000 and 8000 cm^3/min. For small oxidant flow rates, the average particle size is just a bit lower with O_2:N_2 than with pure O_2. Flames with oxidant flow rates of less than 2000 cm^3/min and more than 8000 cm^3/min could not be stabilized and for this reason powder production experiments were not carried out for these configurations.

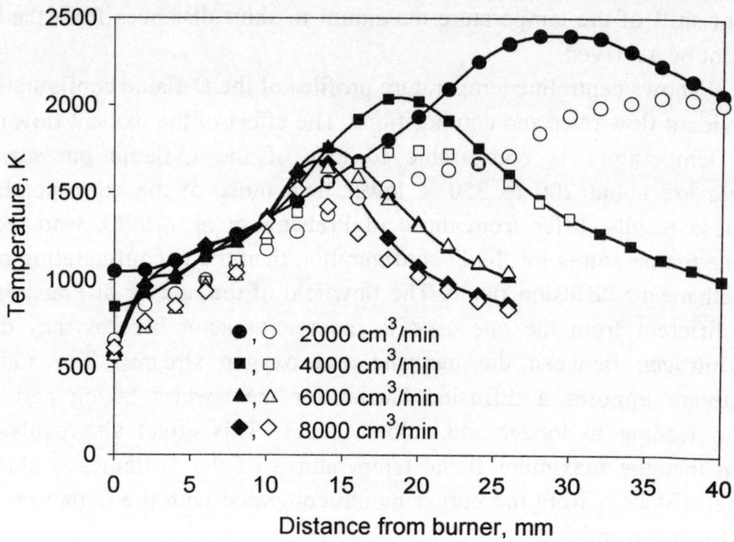

Figure 7. **Temperature profiles of the D-flame measured at the flame centreline in the absence of precursor for pure oxygen flow rates of 2000, 4000, 6000, and 8000 cm^3/min (filled symbols) and an oxidant composition of O_2:N_2=2:1 (open symbols).**

The results of nanoparticle synthesis in the D-flame are depicted in Figure 9 giving the specific surface area as a function of the oxidant flow rate for the two oxidant compositions. For pure oxygen as oxidant (circles) the increase in SSA with the oxidant flow rate is similar to that observed for the C-flame, however the maximum SSA of about 240 m^2/g (corresponding BET-equivalent particle diameter: 11 nm) reached at 10000 cm^3/min, exceeds that of the C-flame by 40 m^2/g. On the other hand, the D-flame gives smaller particles for oxygen flow rates below 8000 cm^3/min. Similar findings were made for O_2:N_2 = 2:1 as oxidant (squares in Figures 8 and 9).

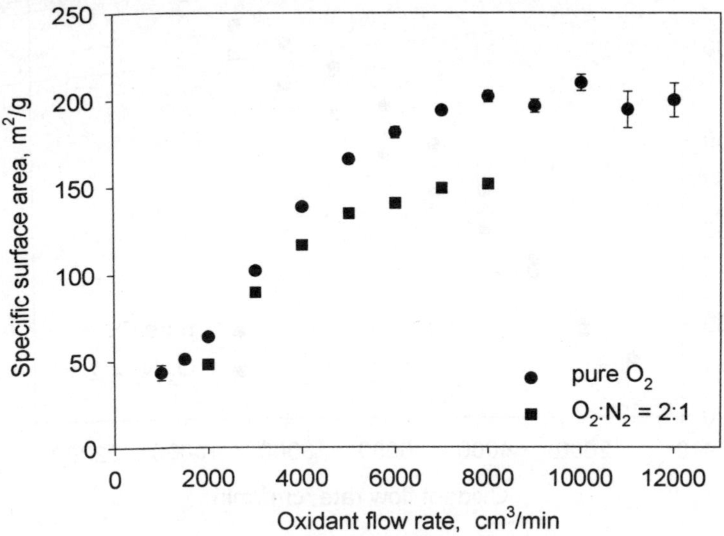

Figure 8. Influence of the total oxidant flow rate on the specific surface area of silica particles synthesized in the C-flame with two different oxidant compositions.

The temperature history and the residence time of the particles in the flame are known to be the main factors influencing the particle growth mechanism (Pratsinis, 1998). The residence time of the particles in the hot reaction zone of the flame depends on the gas flow rates of the reactor in two ways: first the region of particle growth is shorter at high gas flow rates since the flame temperature decreases faster when the oxidant flow rate is increased (Figure 6), and secondly the particles move faster if the velocity of the gas is increased which further decreases the residence time. Both effects favour the formation of smaller particles. Provided the temperature of the gas is lower than about 2000 K, silica nanoparticles have partly sintered, branchy structures, as was first demonstrated by Ulrich and Subramanian (1977).

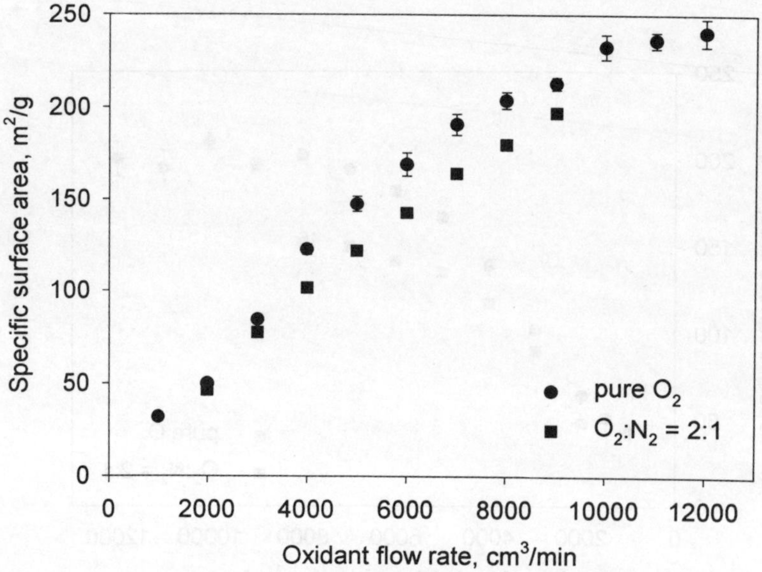

Figure 9. The effect of the oxidant flow rate on the specific surface area of silica particles synthesized in the D-flame with two oxidant compositions.

For temperatures above 2000 K the particle morphology depends on the collision frequency as was shown by Ulrich and Subramanian (1977) and Ulrich and Riehl (1982). At high number concentrations silica particles collide more rapidly than they can fuse, whereas at low number concentrations complete coalescence of the fractal-like aggregates can take place and spherical nanoparticles can be formed: thus the residence time of the particles at high temperatures strongly affects the morphology of the product powders. At oxidant flow rates of 6000 cm^3/min and higher, leading to lower temperatures, the melting point is just reached or exceeded for a short period of time. Resulting product particles are aggregates of small primary particles as can be seen from TEM micrographs presented in Figures 10c and d.

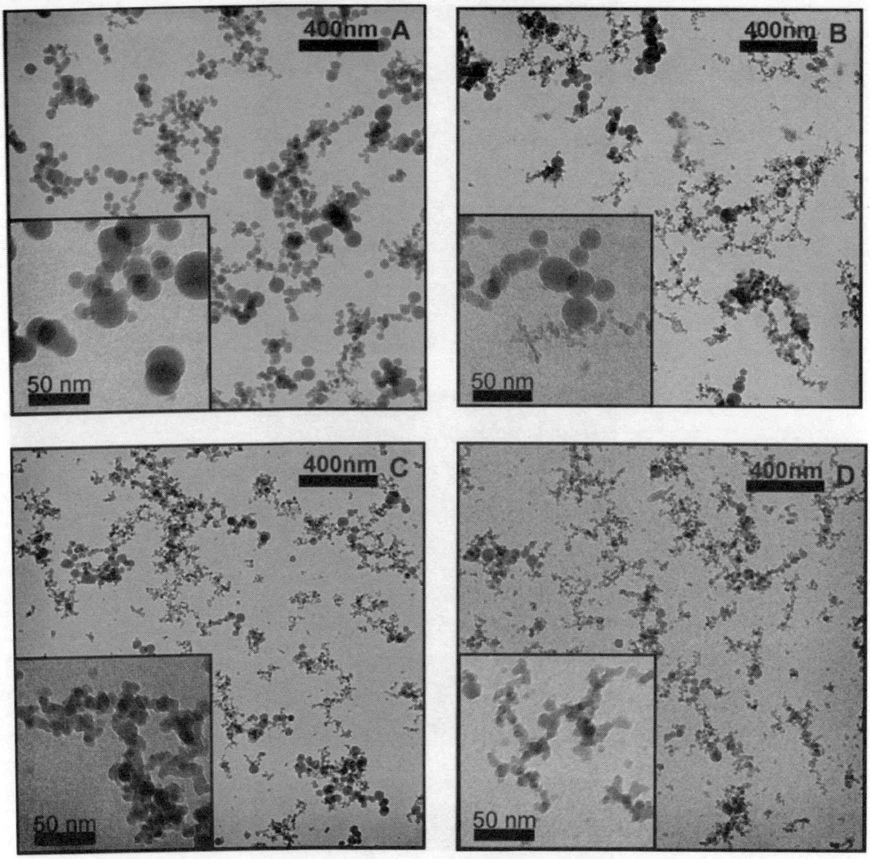

Figure 10. TEM micrographs of silica nanoparticles synthesized in the C-flame using pure oxygen at flow rates of a) 2000, b) 4000, c) 6000 and d) 8000 cm^3/min.

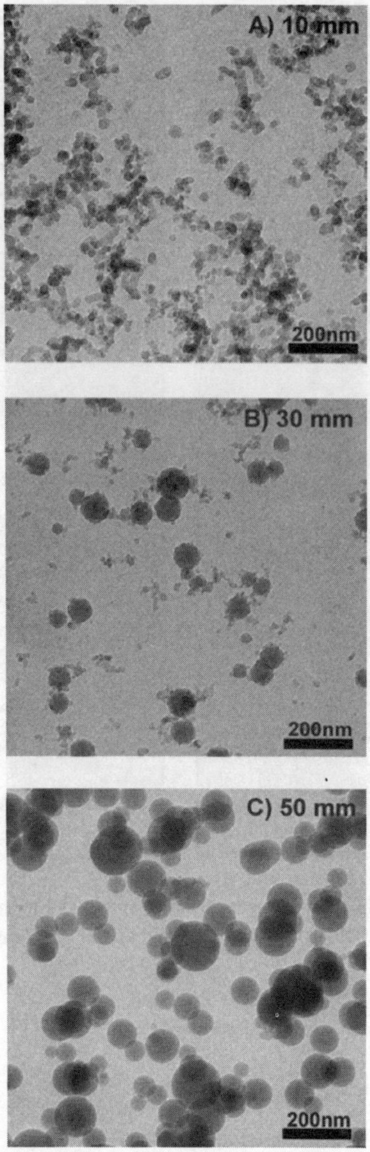

Figure 11. TEM micrographs of fumed silica obtained by thermophoretic sampling in the C-flame (oxygen flow rate: 2000 cm^3/min) at a) 10 mm, b) 30 mm and c) 50 mm from the burner mouth. Particles grow from fractal-like aggregates at 10 mm to spherical particles at 50 mm.

Particles synthesized in flames with lower oxidant flow rates stay longer in the temperature region above 2000 K (Fig. 6). This leads to the formation of a large number of rather big spherical particles with diameters of 50 to 100 nm and a low degree of aggregation at an oxygen flow rate of 2000 cm^3/min, as can be seen in Figure 10a. Aggregates of small primary particles (dp $\approx$ 30 nm) are also found, indicating that the particles have different temperature histories arising from temperature gradients perpendicular to the centreline of the flame. Particle growth in this flame can be illustrated by thermophoretic sampling at 10, 30 and 50 mm from the burner mouth (Figure 11). At the lowest point particles are still aggregated, as can be seen from Figure 11a whereas at 30 mm, some aggregates have fused to larger spherical particles (Fig. 11b) and smaller fractal-like nanoparticles are still present, many of which are attached to the larger spheres. Only spherical particles are found at 50 mm (Fig. 11c) as the small particles in Fig. 11b have completely fused into the large ones, consistent with theory. Sintering necks between the particles cannot be detected in Figure 11c. The large number of aggregates in the product powder (Fig. 10a) indicates however that coagulation and fusion still continues in the upper part of the flame. Only a few spherical particles are found in the product powder when the oxygen flow rate is 4000 cm^3/min (Figure 10b), implying that the residence time at high temperature is still too short for complete coalescence.

In the case of the D-flame the maximum flame temperature is reached at larger distances from the burner mouth, especially for small oxidant flow rates (Fig. 7). As a result, the time taken for silica particles to grow by coagulation into branchy aggregates is longer for the D-flame configuration. When the melting point of silica ($T_{melt} \approx$ 1800 K, Ulrich and Riehl, 1982) is reached at prolonged residence times, fusion continues and the collision frequency diminishes, resulting in particles with lower specific surface area for oxidant flow rates below 8000 cm^3/min for pure O_2 and below 6000 cm^3/min for O_2/N_2 as oxidant. This leads to the formation of mainly non-aggregated and spherical silica particles for an oxygen flow rate of 2000 cm^3/min as is shown in Figure 12. At higher oxidant flow rates the melting point of silica is no longer reached (Fig. 7) and sintering takes place at slower rates resulting in higher specific surface area powders for the D-flame configuration.

The results of nanopowder synthesis with the diffusion flame microreactor can be best compared with Briesen et al. (1998a), as their reactor operates with similar gas flow rates in the C-flame configuration and is also used for the production of silica nanoparticles from hexamethyldisiloxane as precursor. A comparison of the

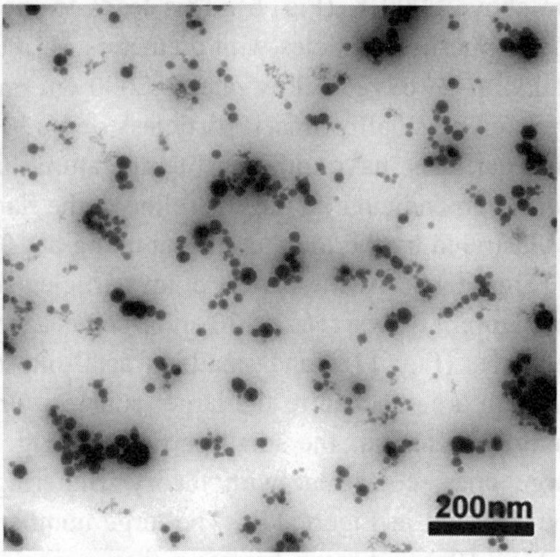

Figure 12. TEM micrograph of spherical silica nanoparticles synthesized in the D-flame at an oxygen flow rate of 2000 cm^3/min.

maximum specific surface area obtained with oxygen as oxidant shows that the microreactor produces much smaller particles. Briesen et al. could synthesize particles with a SSA of about 42 m^2/g at their highest oxygen flow rate of 5500 cm^3/min; at the same flow rate the microreactor generates particles with a specific surface area about four times higher, as can be seen in Figure 8. The difference can be explained by shorter residence times of the particles in the flame of the microreactor and steeper temperature gradients. The small size of the reactor results in outlet velocities 15 times higher for similar volume flow rates and consequently less time is available for particle growth. Furthermore, the velocity ratio between the oxidant and precursor/fuel jets is about 1.5 times higher in the case of the microreactor and with 1.8 mm the inner reactor nozzle diameter is 2.8 times smaller. Thus reactant mixing is enhanced in the case of the microreactor as is demonstrated by a comparison of the cold flow concentration profiles of the two reactors (Figures 4 and 5) as well as temperature measurements by the example of the C-flame configuration with oxygen as oxidant. The maximum centreline temperature of the two flames is reached within a distance of 10 mm from the burner exit, however the maximum flame temperatures of the microreactor are up to 400 K higher and the cooling gradient following the maximum is steeper. The flames of the microreactor reach a temperature of 1000 K within a distance of 25 mm. The same temperature was measured at a height of 40 mm in the reactor of Briesen et al. so that particles are exposed to high temperatures for a longer time leading to powders with lower specific surface area.

The microreactor is also more sensitive to an increase of the oxygen flow rate than the larger reactor of Briesen et al. (1998a), allowing a broader operation and giving rise to greater flexibility in product quality. An increase in oxygen flow from 2000 to 6000 cm^3/min leads to a 2.8fold increase in the specific surface area. Briesen et al. (1998a) observed only an increase by 1.2 for similar oxygen flow rates. A broader range of temperature profiles of the microreactor, expressed by the shift of the maximum temperature from the burner mouth accounts for this observation.

4. Conclusions

A co-flow diffusion flame reactor with a diameter of less than 1 cm was developed and performance-tested for synthesis of silica nanoparticle. The influence of the oxidant flow rate on the nanoparticle properties was investigated for two flame configurations and oxidant compositions. The specific surface area of the powders produced could be precisely controlled from 40 to 200 m^2/g in the case of the C-flame configuration and between 30 and 240 m^2/g in the case of the D-flame configuration when oxygen was used as oxidant. The use of an oxygen / nitrogen mixture in the molar ratio of 2/1 as oxidant gives specific surface areas as much as 50 m^2/g lower. Spherical and non-aggregated silica particles could be synthesized in the D-flame at an oxygen flow rate of 2000 cm^3/min. The process of coagulation and sintering was shown by thermophoretic sampling as a function of flame height. The small-scale diffusion flame reactor exhibits improved properties generally attributed to microreactor systems. This was shown, for example, by enhanced reactant mixing in the flame as was indicated by the measured flame temperature and calculated concentration profiles. The microreactor is more sensitive to changes in oxidant flow rates than a larger scale reactor operated with similar reactant flows. This allows a broader field of operation, making it possible to produce high surface area, fractal-like particles as well as spherical ones simply by changing the oxidant flow rate.

Acknowledgments

Financial support by ETH Gesuch Nr. 33/99-2 is acknowledged.

References

1. Arkles B., 1995, Silicon, Germanium, Tin and Lead Compounds, Metal Alkoxides, Diketones and Carboxylates (Crelest Inc., Tallyton, PA).
2. Briesen H., *et al.*, 1998a, *Chem. Eng. Sci.* **53**, 4105-4112.
3. Briesen H. *et al.*, 1998b, in *Mat. Res. Soc. Symp. Proc.*, eds. Beaucage G., Mark J.E., Burns G.T., Hua D.W (MRS, Warrendale), 3-14.
4. Burns J.R. et al., 1997, in Conference Reprints 2nd Intl. Conf. on Microreaction Technology, New Orleans, 39-44.
5. Chigier N.A. and Béer J.M., 1964, *Trans. A.S.M.E., J. Basic Eng.* **86**, 797-805.
7. Dobbins R.A. and Megaridis C.M., 1987, *Langmuir* **3**, 254-259.
8. Ehrfeld W. et al., 1997, in Abstracts 1st Intl. Conf. on Microreaction Technology. (DECHEMA e.V., 1997), 9/1-9/2.
9. Fotou G.P. *et al.*, 1995, *Combustion and Flame* **101**, 529-538.
10. Günther R., 1974, *Verbrennung und Feuerungen* (Springer Verlag, Berlin), 125-192.
11. Glumac N.G. *et al.*, 1998, *Mater. Letters* **34**, 148-153.
12. Johannessen T. *et al.*, 2000, *Chem. Eng. Sci.* **55**, 177-191.
13. Kammler H.K. and Pratsinis S.E., 1999, *J. Nanoparticle Res.* **1**, 467-477.
14. Kammler H.K. and Pratsinis S.E., 2000, *Chem. Eng. Process.* **39**, 219-227
15. Ko N.W.M. and Kwan A.S.H., 1976, *J. Fluid Mech.* **73**, 305-332.
16. Lide, D.R., 1997, CRC Handbook of Chemistry and Physics 78th ed. (CRC Press, Boca Raton).
17. Livage J. *et al.*, 1988, *Prog. Solid State Chem.* **18**, 259-341.
18. Pratsinis S.E. *et al.*, 1996, *Powder Technology* **86**, 87-93.
19. Pratsinis S.E., 1998, *Prog. Energy Combust. Sci.* **24**, 197-219.
20. Pratsinis S.E. and Mastrangelo S.V.R., 1989, *Chem. Eng. Prog.* **85**, 62-66.
21. Talamelli G.B.A. and Petagna P., 1994, *Exp. Thermal and Fluid Sci.* **9**, 135-146.
22. Ulrich G.D. and Subramanian N.S., 1977, *Comb. Sci. Techn.* **17**, 119-126.
23. Ulrich G.D. and Riehl J.W., 1982, *J. Colloid Interface Sci.* **87**, 257-265.

24. Ulrich G.D., 1984, *Chem. Eng. News* **62**, 22-29.
25. Vemury S. and Pratsinis S.E., 1995a, *J. Amer. Ceram. Soc.* **78**, 2984-2992.
26. Vemury S. and Pratsinis S.E., 1995b, *Appl. Phys. Lett.* **66**, 3275-3277.
27. Villermaux E. *et al.*, 1994, *Meccanica* **29**, 393-401.
28. Wegeng R.S. et al., 1996, in Proceedings of the 1996 Spring National Meeting of the American Institute of Chemical Engineers, New Orleans.
29. Wegeng R.S. and Drost M.K., 1998, in Conference Reprints 2nd Intl. Conf. on Microreaction Technology, New Orleans, 1998, 3-9.
30. Wießmeier G. and Hönicke D., 1996, *J. Micromech. Microeng.* **6**, 285-289.
31. Wooldridge M.S., 1998, *Prog. Energy Combust. Sci.* **24**, 63-87.
32. Zachariah M.R. and Semerjian H.G., 1990, *High Temp. Sci.* **28**, 113-125.
33. Zhu W. and Pratsinis S.E., 1996, in *Nanotechnology: Molecularly Designed Materials*, eds. Chow G.M. and Gonsalves K.E. (ACS Symp. Ser. **622**), 64-78.
34. Zhu W. and Pratsinis S.E., 1997, *AIChE Journal* **43**, 2657-2664.

CHAPTER EIGHT

SYNTHESIS OF NANOPOWDERS BY THE MICROWAVE PLASMA PROCESS - BASIC CONSIDERATIONS AND PERSPECTIVES FOR SCALING UP

Dieter Vollath, D. Vinga Szabó

Forschungszentrum Karlsruhe, Institut für Materialforschung III
P.O.Box 3640, D-76021 Karlsruhe, GERMANY
E-mail: dieter.vollath@imf.fzk.de

1. Introduction

In general, nanomaterials are defined as materials with grain sizes below 100 nm. A more stringent definition might be that nanomaterials exhibit special properties which are dependent on their small grain size. In many cases, the latter definition restricts nanomaterials to grain sizes below 10 nm.

Nanocrystalline ceramic materials or nanoglasses with particle sizes below 10 nm exhibit interesting physical properties [1] which, apart from those related to grain boundaries, are those of single isolated particles. These properties however are lost where an interaction between the particles is possible and thus such interaction must be minimised by keeping them a certain distance apart. This distance may be smaller in the case of tunnelling interaction, or larger in the case of dipole-dipole interaction which leads to the concept that nanocomposites consist of two phases: an active phase carrying the physical property of interest, and a passive phase maintaining distance. It is crucial in producing nanocomposites to establish a proper balance between these two phases.

In general, there are two approaches to producing uniform nanocomposites:

- Synthesis of a metastable solution and precipitation of the second phase by change in temperature.[2]
- Synthesis of each phase separately, blending during the stage of particle formation. Coating the particles of a first reaction stage with the second phase leads to a product with the greatest uniformity.

Processes based on mechanical blending never lead to homogeneous products on the nanometer scale. The only way to obtain the starting material for a uniform nanocomposite of different phases is to coat the particles of the active phase with the passive phase. [3]

Gleiter et al. [4,5] and Siegel et al. [6] developed the standard process, the inert gas condensation method, for synthesis of nanosized metal or oxide powder. Variations of this process utilise an electron beam [7] for the production of the metal vapour or laser ablation [8], evaporating the material directly as an oxide. Synthesis of superfine oxide powders is also possible using flame hydrolysis and hydrothermal synthesis [9]. Both flame hydrolysis and hydrothermal synthesis, and the laser ablation process lead to highly agglomerated powders.

The gas condensation process [4-6] was the first process to produce non-agglomerated nanoscaled particles. In this process, a metal is evaporated in an inert atmosphere at relatively low pressure; the metal vapour loses its high thermal energy by random collisions in the gas atmosphere and forms nanoparticles. These metal particles migrate by diffusion in a temperature gradient to a liquid nitrogen-cooled cold finger. To obtain oxide particles the metal particles collected on the surface of the cold finger are oxidised by adding oxygen to the atmosphere. The reaction product collected on the cold finger is scraped off mechanically.

The following processes take place to form the particles in this process:

- The energy of the metal vapour atoms is reduced to the thermal mean value by collisions with inert gas atoms.
- The collision of two or more metal atoms in the inert atmosphere forms nuclei.
- Additional metal atoms condense on the surface of these nuclei leading to an increase in the particle size.
- Collisions of these particles lead to coagulation by exchange of surface energy.
- If the size of the particles formed by coagulation exceeds a certain size, additional collisions with other particles lead to agglomerates. As long as this process occurs at relatively low temperatures the agglomerates are kept together by van der Waals bonds. At higher temperatures however the particles sinter, forming hard agglomerates.

To produce monocrystalline nanoparticles, conditions favourable for agglomeration must be avoided. A low density of particles in the gas atmosphere reduces the production rate significantly and is the only way to achieve this goal. The inert gas condensation process has two major drawbacks: small production capacity and a narrow range of compounds which can be synthesised as nanoscaled powder. The microwave plasma method for the synthesis of nanocrystalline ceramic powders overcomes both these problems for the first time. This process utilises chemical reactions in a plasma environment. [10-13] The main advantages of the microwave process are low reaction temperature and conditions for synthesis which are not favourable for particle agglomeration. This process will be explained in detail in this Chapter.

A later variation of this process uses a tubular furnace [14] instead of a microwave plasma. This process is also able to achieve a broad range of chemical reactions with increased variability of precursors and products. In most cases the reaction temperature is in the range of ca. 1000 °C. This high temperature is necessary to overcome kinetic obstacles to the chemical reactions and to increase the reaction rate. To obtain a comparable temperature in a streaming gas, a long heated zone is required which leads to a longer residence time of the particles in the hot zone. Both high temperature and long residence time are essential for the reaction kinetics and high reaction rates. A high reaction temperature however increases the probability of formation of sintered agglomerates and a long residence time at high temperature broadens the particle size distribution significantly: both these phenomena lead to poor quality products.

The common characteristics of the microwave plasma process and the reaction in a tubular furnace are compared with the inert gas condensation processes as follows:

- The intended reaction product is already formed in the nucleation stage and not in a subsequent reaction.
- The range of reaction products is not - at least theoretically - limited. In most cases, the reaction products have higher melting points than the pure metals which allows the potential for a reduced probability for sintered agglomerates.

The particles produced are collected either by thermophoresis or by filtration systems. The application of filtration systems in laboratory equipment is problematic because it is difficult to keep the system at constant pressure during the fabrication process. Large industrial plants may use filtration systems to separate the product from the gas stream more easily.

2. The Microwave Plasma Process

2.1 Basic considerations

Obviously high production rate, unagglomerated particles, and narrow particle size distribution are at odds with the classical processes of gas phase synthesis. A high production rate is obtained only with high particle density in the gas stream and a high particle density provokes particle agglomeration. To avoid sintering of the agglomerates, it is necessary to work at the lowest possible temperature where kinetic obstacles prevent the chemical reactions. To obtain the necessary high temperature in the fast flowing reaction gas, a relatively long residence time at elevated temperature is necessary and this leads to a broad particle size distribution.

The microwave plasma process solves all these problems. Dissociation and ionisation of the reactants in the plasma remove the kinetic obstacles which reduces the reaction temperature significantly. Furthermore, ionised reactants lead to ionised particles and as the particles carry electrical charges with the same sign, they repel each other, avoiding extensive particle growth and agglomeration.

To clarify the advantages of the process presented in this paper, it is necessary to understand the physics of a microwave plasma in which the temperatures are significantly lower than a dc or ac plasma. This is because the energy E transferred in an oscillating electrical field is inversely proportional to the mass of the particle m and the squared frequency f.

$$E \propto \frac{1}{mf^2} \tag{1}$$

As the mass of the electrons is small in comparison with the mass of the ions, at high frequencies a substantially larger amount of energy is transferred to the electrons, than to the ions. Apart from ions and free electrons, microwave plasma consists of neutral gas species, dissociated gas and precursor molecules. Collisions between charged and uncharged particles therefore influence the transfer of energy to the particles and the collision frequency z has thus to be considered:

$$E \propto \frac{1}{m} \frac{z}{f^2 + z^2} \tag{2}$$

This does not alter the mass relationship of the energy transfer but it introduces an element of dependency of the collision frequency in the plasma according to the gas pressure. Thus, the energy of the free electrons, the "temperature", is significantly higher than the "temperature" of the ions [15] and the “temperature” of the neutral gas molecules, obtaining their energy from collisions with the charged molecules, is even lower. Consequently the "overall temperature" of a gas passing a microwave plasma is not as high as that of a dc or rf plasma, where temperatures in the range from 5000 to 15000 °C are obtained. According to formula (2) the gas pressure has a significant influence on the synthesis as the collisions between energy-rich charged particles and uncharged particles control the energy transfer. The collision frequency z is proportional to the gas pressure. Analysing formula (2) one learns that for $f < z$ the energy transfer increases, whereas for $f > z$ the energy transfer decreases with increasing gas pressure. Thus, depending on the gas pressure and the input of microwave energy, the temperature determined directly after the plasma zone can be adjusted in a range from 150 to 900 °C. Gas pressure and energy input are used to achieve optimal conditions for synthesis for the desired chemical reactions and the product.

In contrast to a conventional synthesis in a tubular furnace, the mechanism of the formation of the nanoparticles in a microwave plasma is assumed to occur in the following steps: [3,16]

- Ionisation and dissociation of the reactive components of the plasma.
- Reaction of the dissociated species forming molecules of the intended compound.
- Homogenous nucleation of the particles by random collision of two or more molecules.
- Growth of the nuclei by further collision with molecules.
- Further growth by coagulation of the particles: the increase of the electrical charge of the particles with size limits this coagulation process.

The particles leave the reaction zone are non-agglomerated with electrical charges of the same polarity. This makes it possible to coat each individual particle in a second reaction step.

In the reaction zone, a gaseous precursor reacts with a reaction gas for instance, in the case of an oxide a mixture of argon with oxygen is used as a reaction gas. The central problem with this kind of synthesis is the selection of an appropriate precursor. If possible, water-free chlorides or carbonyls are selected as precursors and in cases where these compounds do not exist or are not sufficiently volatile, metal-organic compounds are used. Assuming a chloride as precursor, the following reactions are assumed in the plasma:

$$MeCl_n + \frac{m}{2}O_2 \Rightarrow MeO_m + \frac{n}{2}Cl_2 \tag{3}$$

Adding water shifts the equilibrium more to the right hand side. In this case, in the plasma water dissociates and forms H^+ and OH^- radicals making two reaction routes possible:

- i) $$MeCl_n + \frac{m}{2}O_2 + \frac{n}{2}H_2 \Rightarrow MeO_m + nHCl \tag{4}$$

The hydrogen stems from the recombination of the protons with free electrons from the plasma.

$$2H^{+} + 2e^{-} \Rightarrow H_2 \tag{4a}$$

- ii)
$$\begin{aligned} &MeCl_n + (m+x)OH^{-} + (m+x)H^{+} \Rightarrow \\ &MeO_m + xHClO + (n-x)HCl + (m+2x-n)H_2 \end{aligned} \tag{5}$$

Where $x << n$ is assumed.

The addition of water has two effects: through the formation of HCl, and possibly $HClO$, it increases the reaction enthalpy and acts as a catalyst in increasing dissociation. The temperature of the gas, monitored after passing the reaction (plasma) zone, is generally higher in the case of added water and water additions also change the electric charges of the particles, inhibiting particle growth. In summary, the positively charged particles are neutralised by collision with OH^{-} ions which quenches the mechanism limiting particle growth.

To synthesise nitrides, nitrogen is used as reaction gas. As chlorides are more stable than nitrides, it is necessary to add hydrogen to the system to shift the equilibrium to the side of the nitride and HCl formation.

$$MeCl_n + \frac{m}{2}N_2 + \frac{n}{2}H_2 \Rightarrow MeN_m + nHCl \tag{6}$$

Although in many cases there is the problem of the formation of NH_4Cl, it is beneficial to use ammonia as a hydrogen carrier.

Using similar reactions it is also possible to synthesise sulphide, selenide, carbide and, in particular cases, metallic nanoparticles.

As the particles leave the reaction zone with electrical charges preventing agglomeration, it is possible to extend the process to produce coated nanoparticles. [3] In this context, one may think of, for instance, the synthesis of oxide particles consisting of a core MaO_n coated with a second oxide MbO_m. Two identical consecutive plasma stages on the reaction tube, each with its own supply of reactive gases, are used to produce ceramic-coated nanoparticles. In the second plasma stage, the chemical reactions are the same as in the first reaction stage in which the particles are formed by homogenous nucleation. In contrast, in the second reaction stage the particles formed in the first step act as heterogeneous nuclei for the molecules of the second oxide MbO_m. The gas kinetic collision cross-section of the particles formed in the first step is about two orders of magnitude larger than the cross section of a molecule and the probability of the condensation of *Mb* oxide molecules on the *Ma* oxide particles is thus also at least two orders of magnitude larger: consequently one would expect a high yield in coated particles. Based on these considerations, it is obvious that a small amount of pure particles of *Mb* oxide is unavoidable; it is however impossible for any of the *Ma* oxide particles to remain uncoated.

2.2 The experimental set-up

The experimental set-up for the synthesis of ceramic nanopowders can be realised with industrial standard frequencies 0.915 and 2.45 GHz. In the case of the higher frequency, the maximum diameter of the reaction tube is ca. 30 mm, whereas it is possible to design a system with an 80 mm reaction tube using a 0.915 GHz system. For synthesis of larger amounts of nanopowders therefore, the use of 0.915 GHz is generally advantageous. Apart from the question of quantity, it is important to note that because of the differences in energy transfer (see formulae (1) and (2)) the application of 2.45 GHz systems is generally advisable in cases where low temperatures are crucial.

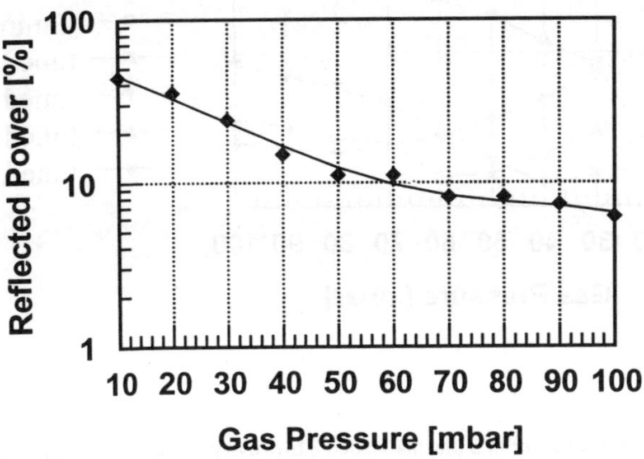

Figure 1. Reflected power in a TE_{01} cavity tuned for ignition.

In the case of a 0.915 GHz system, the reaction is performed in a 50 mm diameter reaction vessel made of quartz, passing a TE_{01} or TE_{11} mode cavity connected to a microwave generator. [17] The TE_{01} microwave cavity, made of aluminium, is designed as a part of a single-mode microwave system using an IEC 153 R9 (equivalent to RETMA WR 975) waveguide. A maximum of the microwave E-field vector at the centreline of the reaction tube is adjusted using a sliding short and a tri-stub tuner. The set-up for the TE_{11} cavity, usually made of stainless steel, is essentially the same apart from the fact that a sliding short is not necessary. The latter cavity has - as shown in figures 1 and 2 - a better microwave efficiency compared with that of the TE_{01} cavity. In general, this may lead to higher temperatures in the reaction zone. Figures 1 and 2 show the reflected power of the plasma-loaded cavities as a function of the gas pressure and the tuning. The fact that the TE_{01} cavity can not be tuned for optimum operating conditions significantly reduces the efficiency of the microwaves. Also, figure 2 shows that for the TE_{11} cavity the region of efficient operation is relatively broad: for instance, a system tuned for 40 mbar has good operating conditions in the range from 10 to 70 mbar and such a cavity allows the application of a larger diameter of the reaction tube. This is important for the scaling-up of the process.

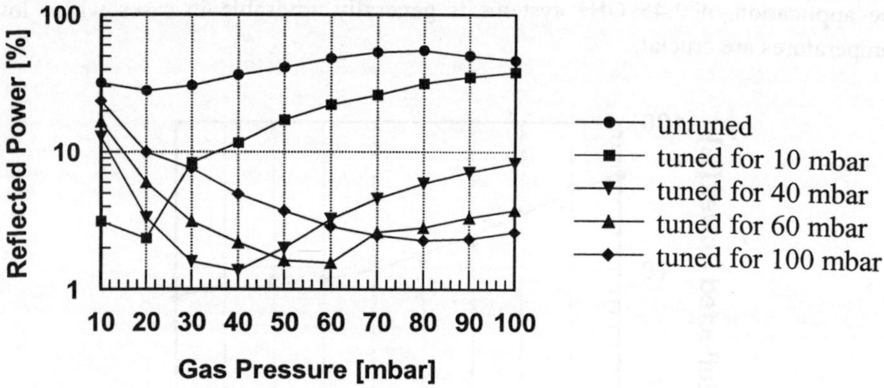

Figure 2. Reflected power in a plasma loaded TE_{11} cavity tuned for different gas pressures.

To obtain coated nanoparticles, this set-up uses two plasma zones placed consecutively on one reaction tube. Figure 3 shows a schematic view of the arrangement with TE_{01} cavities. Microwaves generated in a magnetron pass an isolator and the waveguide is split into two branches using an E-plane power splitter or 3dB divider. Each of the cavities is situated between a tri-stub tuner and a movable short and each branch is equipped with a directional coupler to measure the energy of the forward and backward wave: isolators are used to decouple the two branches since without this it is extremely difficult to come to a stable operating plasma in both branches. The TE_{01} cavity can be operated only in a tuning optimal for plasma ignition whereas the plasma in the TE_{11} cavity can be ignited with the same tuning optimal for operation.

In the actual set-up of this process, the vapours of the precursors in question are produced outside the reaction tube and are introduced in front of the plasma zones. This makes it possible to produce coated nanoparticles, a two-phase composite. Preheating of the gas carrying the precursor is necessary to avoid precipitation which leads to large particle sizes. Complete evaporation of added water is important before it enters the zone where the vaporised precursor is introduced, otherwise the droplets act as condensation nuclei for the precursor, leading to very large particle sizes. As a next step, the gas mixture consisting of the carrier gas, the vaporised precursor, and eventually water vapour, passes the plasma zone where the oxide or nitride nanoparticles are formed.

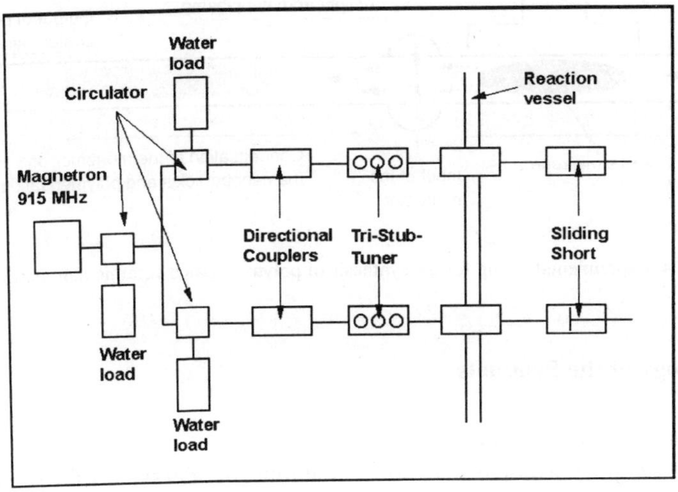

Figure 3. Schematic drawing of the arrangement to produce coated nanoparticles using TE_{01} cavities.

In a typical run to produce coated oxide nanoparticles, a gas volume of about 75 l/min STP passes the plasma zone. A mixture of argon with 20 vol% oxygen is used as plasma gas which carries the evaporated precursor. The pressure in the reaction tube is selected from within the range 10 to 50 mbar. The temperature is always kept below 600 °C; keeping the temperature at a low level is difficult because both plasma sources introduce energy into the system.

In many instances, the final product, a sintered body, is not exposed to any severe thermal or mechanical load and polymer binding between the particles is therefore sufficient. This leads to polymer coated particles. [18,19] In this case, the polymer acts as distance holder as well as a binder. Figure 4 depicts the equipment used to produce polymer coated nanoparticles. In this case, behind the reaction zone, the vapour of a monomer is introduced into the system. The particles leaving the reaction zone act as condensation seeds for this vapour, provided temperature and monomer are selected properly. The condensate polymerises instantly under the influence of temperature and the uv-radiation from the plasma.

The temperature of the gas is determined after passing the reaction (plasma) zones and cooled surfaces collect the powder by thermophoresis.

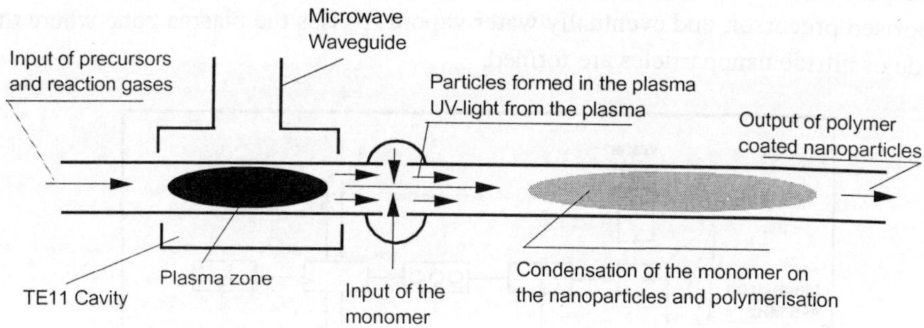

Figure 4. Experimental set-up for the synthesis of polymer coated ceramic nanoparticles.

3. Morphology of the Products

3.1 Single phase ceramic nanopowders

Single-phase ceramic nanoparticles are crystalline or glassy. Surface energy and free enthalpy of melting control crystallisation [21,21] and thus for each ceramic material a critical particle size exists which will control the transition from glassy to crystalline.

Based on these considerations, the transition from glassy to crystalline for zirconia, ZrO_2, takes place in particle sizes below 1 nm. Figure 5 shows a typical electron micrograph of nanocrystalline zirconia. This micrograph reveals nearly spherical particles with a narrow size distribution if about 5 nm. Dark field imaging and high-resolution electron microscopy show that the individual particles are single crystals free of dislocations or twinning. As expected there are no glassy particles.

By electron diffraction, a structure close to the cubic fluorite lattice is identified. At a first glance, this follows Ostwald's rule that the high temperature modification takes place first, however a detailed study using perturbed angular correlation measurements reveals that in this structure the Zr^{4+} cations are in their regular lattice positions, whereas the O^{2-} anions are distributed randomly in the vicinity of their lattice places. [22] Reducing the feed rate of the precursor leads to smaller particles which is demonstrated in the particle size spectrum determined by particle mass spectrometry [23-27] shown in figure 6.

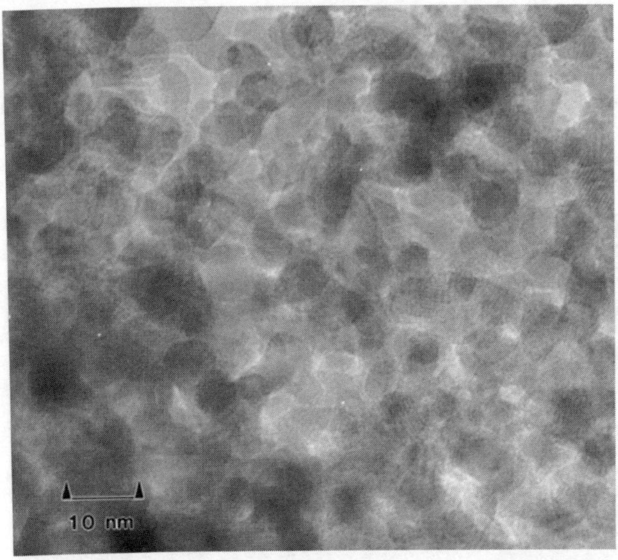

Figure 5. Electron micrograph of zirconia particles

The zirconia examples shown had narrow particle size distributions. As mentioned in chapter 2.1, water additions break down the size limiting mechanism. Figure 7 depicts titania, TiO_2 particles showing this phenomenon clearly. It is obvious that the addition of water leads to a dramatic increase of the particle size and particles with sizes of about 20 to 50 nm are found. In addition it should be noted that the particle size distribution is no longer narrow, however the particles maintain monocrystallinity.

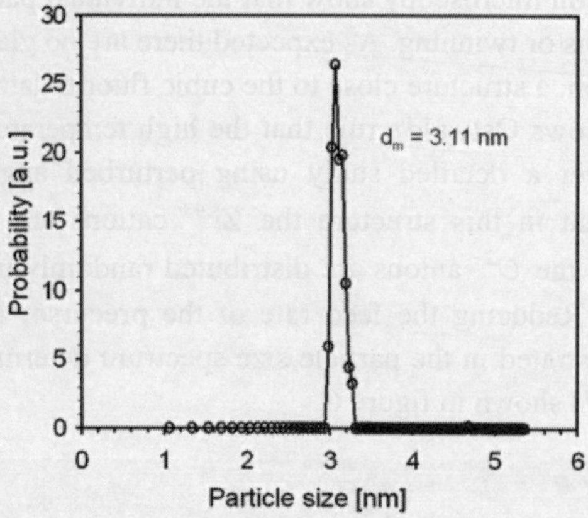

Figure 6. Particle size distribution of zirconia determined by particle mass spectrometry. [23]

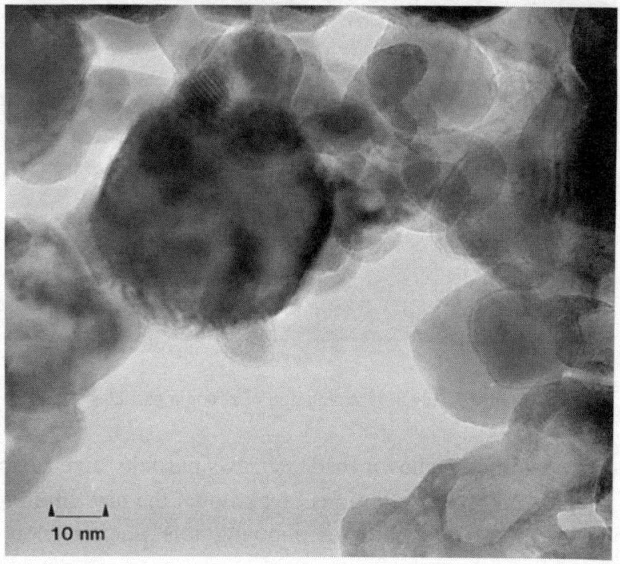

Figure 7. Titania particles synthesised with water additions. The large size of the particles connected to a broad size distribution is characteristic for material synthesised with water additions.

The same narrow particle size distributions as are found in oxides are also typical for nitrides. Figure 8 shows a typical view of zirconium nitride, ZrN, as it is synthesised by this process.

In the examples shown above even the smallest particles are crystallised. This is not so with alumina: dark field electron micrographs reveal a dominant amount of glassy alumina and show some speckles, indicating a few isolated crystallised particles. Electron diffraction patterns exhibit a large amount of diffuse scattering stemming from the amorphous alumina in the specimen, including some intensity resulting from the amorphous carbon film. Finally in the diffraction patterns there is some small indication of crystallised alumina. The cubic γ- alumina fits best with the diffraction patterns. High Resolution Electron Microscopy reveals that particles with sizes below 7 or 8 nm are amorphous and rounded, whereas larger particles are crystalline and faceted. [16] These results indicate a critical diameter for crystallisation of alumina. [16, 28] Figure 9 depicts a typical dark field electron micrograph of nanoparticulate alumina showing this phenomenon.

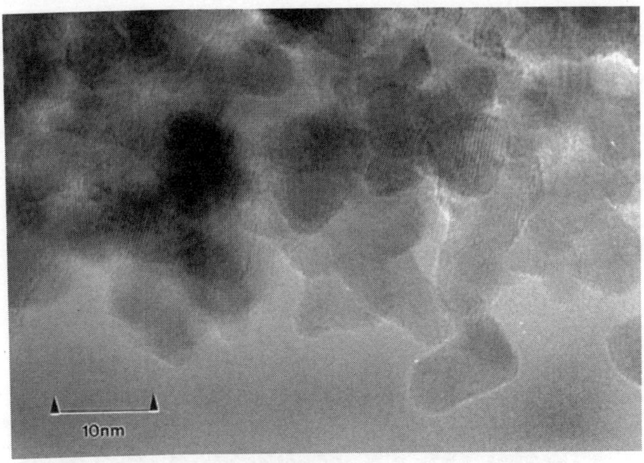

Figure 8. Zirconium nitride particles. The small particle size with its narrow size distribution is controlled by the same charge mechanisms as in the case of oxides.

Examples of less common nanoparticles are those of sulphides and selenides. In this class of compounds, those crystallising in layered structures are of special interest. These compounds are characterised by covalent bindings within the layers and van der Waals binding between the layers. In small particles, at the periphery of the layers there is a large number of dangling bonds contributing significantly to the energy of the system. Closing these bonds minimises the free energy of the particles which leads to closed structures like nanotubes, onion, or polyhedron shaped crystals. [29-33] A typical example of an onion-shaped $ZrSe_2$ crystal is given in figure 10. Selenides are synthesised by adding the stoichiometric amount of $SeCl_4$ to the metal precursor.

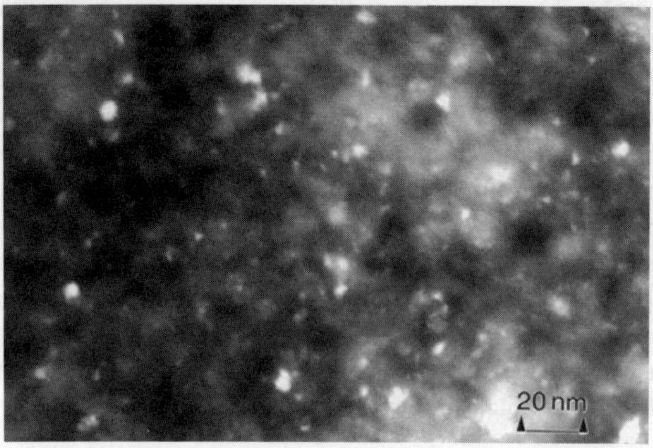

Figure 9. Dark field electron micrograph of nanoparticulate alumina. The bright speckles indicate crystallised particles, whereas the other particles are glassy.

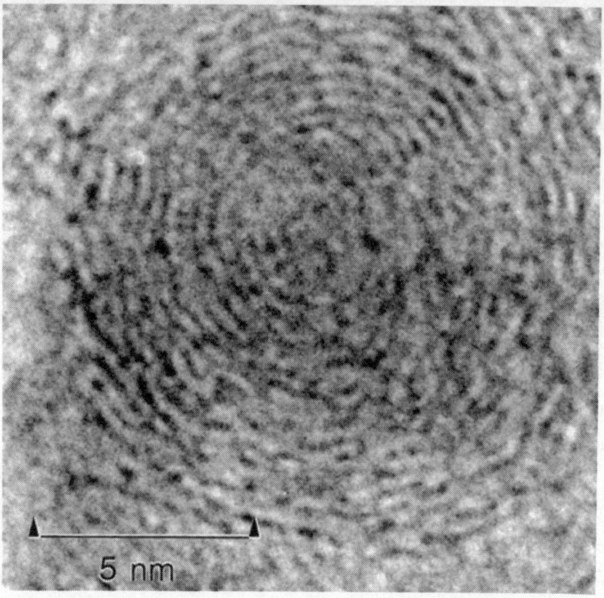

Figure 10. Onion shaped $ZrSe_2$ particle. This material crystallises in closed shapes to minimise the number of dangling bonds at the surface.

3.2 Ceramic coated nanoparticles — ceramic-ceramic nanocomposites

As mentioned in section 2.1, it is possible to make use of the fact that the particles leave the reaction zone unagglomerated with electrical charges to coat the particles individually with a layer of a second ceramic. For thermodynamic reasons, it is possible to coat a particle with a material of the same type; e.g. an oxide core with an oxide coating. It is also a fundamental prerequisite that the phases of core and coating exhibit neither extended mutual solubility nor a compound consisting of the two phases. With respect to crystallisation, the same rules apply as those for bare particles. Depending on the crystallisation behaviour of the two phases forming core and coating, three fundamentally different types of coated particles are formed. The typical structures are given based on the system $Al_2O_3 - ZrO_2$ which is an ideal model for nanocomposite materials because alumina and zirconia exhibit a very low mutual solubility and do not form any compound. [34] The crystallisation behaviour of these two compounds is, as explained in section 3.1 entirely different causing two types of nanocomposites characterised by the existence of one glassy and one crystalline phase. One of these structures depicted in Figure 11 consists of a crystallised core of zirconia characterised by lattice fringes, with a coating of glassy alumina. In this micrograph, the sizes of the cores range from 4 to 6 nm, whereas the coating, visible by a faint halo, is ca. 0.5 nm thick.

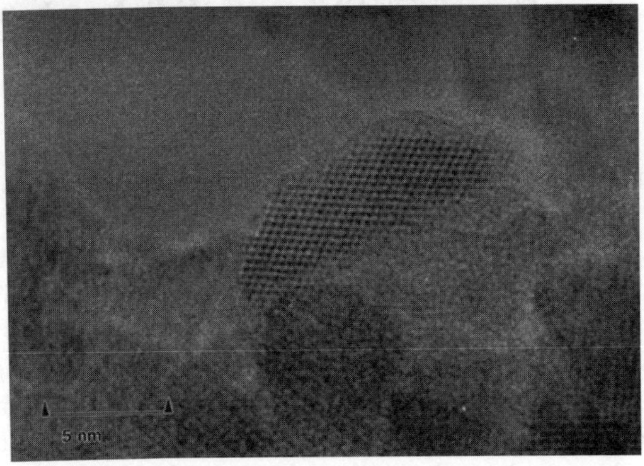

Figure 11. Crystallised zirconia particle coated with ca 0.5 nm glassy alumina.

Additional point analyses of coated ZrO_2 / Al_2O_3 particles using energy dispersive X-ray microanalysis in a dedicated STEM [35] confirmed the interpretation of the electron micrograph. Figure 12 reveals a significantly higher *Al* signal in the coating when compared to the parts closer the core. Since the particle coating is perfect, X-ray signals from the core contain a signal from the coating. On the other hand, the X-ray signal from the coating contains a signal from the core as the electron beam has a finite thickness and a certain drift of the specimen during spectrum acquisition is unavoidable.

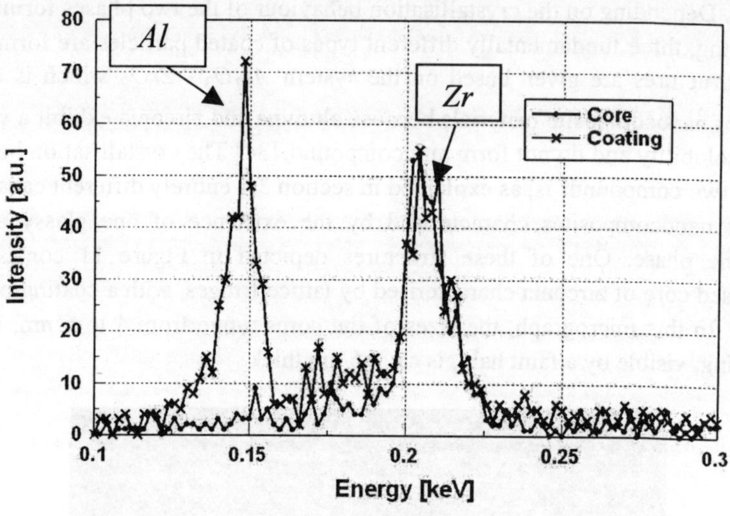

Figure 12. Element distribution in the kernel and coating of a alumina coated zirconia particle.

Coating a glassy alumina core with crystallised zirconia leads to an inverse structure. Electron microscopy depicted in figure 13 shows primarily the lattice fringes of the zirconia coating. In the brighter centre of the particle, the lattice image of the zirconia changes in contrast but not in the image itself, indicating a change in thickness but not in crystal structure. In the micrographs, the centre of these particles is brighter than their outer regions, because of the reduced absorption of the electrons in the amorphous alumina kernel compared with the zirconia coating.

Two crystalline phases are characteristic for the third case. Figure 14 depicts spherical particles with diameters from 6 to 10 nm, consisting of a crystalline zirconia core and a crystalline γ-alumina coating with a thickness of about 2 - 3 nm. The lattice fringes visible in this micrograph represent the {211} fringes of the γ -

alumina of the coating. The lattice fringes contain non-annihilated dislocations. Obviously, these dislocations are sessile ones, necessary to compensate the lattice misfit. Alumina has crystallised epitactically on the lattice of the zirconia kernel. The smallest topotactic misfit [36] of about 10% occurs under the assumption that the alumina has grown on the {111} or {110} plane of the zirconia core. In the example given in figure 14, three additional {211} half planes with opposite directions compensate the topotactic misfit.

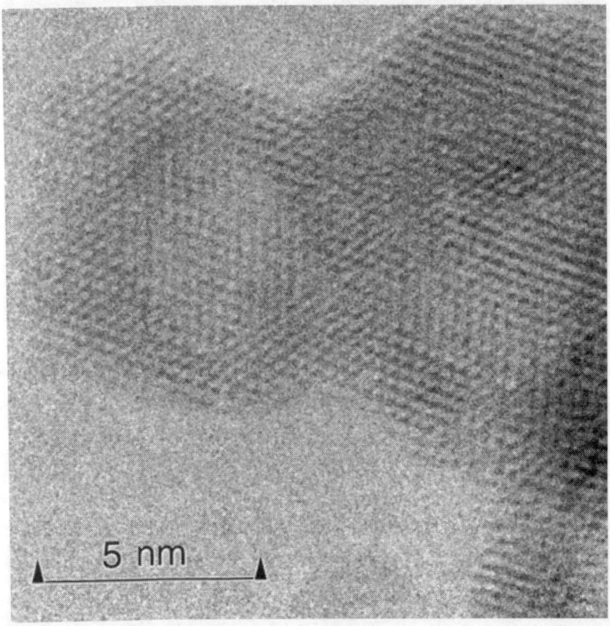

Figure 13. Glassy alumina particle coated with crystallised zirconia. The lattice fringes belong to the zirconia phase.

Pressing and sintering of a powder consisting of zirconia kernels coated with alumina leads to a product with a structure depicted in figure 15.

It is important to note that the size of the zirconia particles did not increase significantly during sintering. Because of the slight solubility of Zr in Al_2O_3 there are zirconium rich areas around the zirconia particles. This can be avoided by selecting material combinations without any mutual solubility.

3.3 Polymer coated nanoparticles — polymer-ceramic nanocomposites

Most of the applications of nanomaterials are not in an environment with severe thermal or mechanical load. Therefore, there are many applications, where a poly-

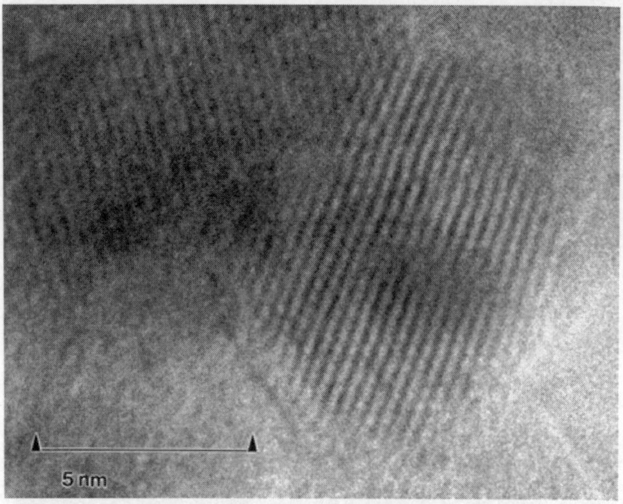

Figure 14. Crystallised zirconia particle coated with crystallised alumina. Misfit dislocations in the alumina lattice, visible in this micrograph, compensate the differences of these two different crystal structures.

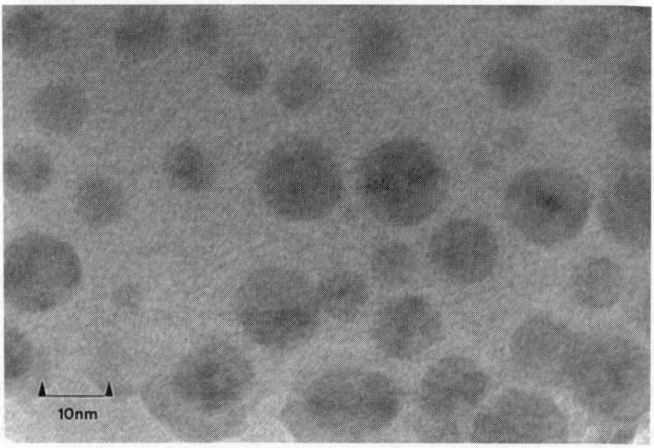

Figure 15. Sintered $ZrO_2 - Al_2O_3$ nanocomposite. It is remarkable to realise that in spite of the full densification, the size of the zirconia particles did not increase during sintering.

mer binding between the particles is sufficient. Polymer coated ceramic nanoparticles are the starting material for these polymer – ceramic nanocomposite. Methacrylic acid (*MA*) and methyl methacrylic acid (*MMA*) are the best monomer for coating ceramic nanoparticles with either *PMA* or *PMMA*. Both monomers polymerise under the influence of temperature and uv-radiation originated from the microwave plasma. Figures 16a,b depict the typical morphology of such a powder synthesised according to the process explained in chapter 2.1. This figure gives in two micrographs an overview and a larger magnification to demonstrate the structure in very detail. Because of the short time, polymerisation is very limited leading to oligomers consisting only of 4 or 5 monomers.

Figure 17 displays the distribution of mol masses of the polymer coating, as it was determined with gel-permeation-chromatography (GPC). Where the mol mass of the monomer is about 100, two dominant peaks of polymerisation arise, one at four and the other at ten monomers. The final polymerisation has to be initiated during densification. Densification of a composite powder, as shown in figure 17, is performed by hot pressing at a temperature in the range from 100 to 150 °C. Under these conditions it is possible to obtain transparent polymer-alumina bodies. Figure 18 shows the morphology of a pressed $\gamma - Fe_2O_3 - PMA$ body: this is an electron micrograph of a specimen prepared by cutting with an ultra microtome. The thickness of the slice was 20 nm. and as this extreme thin specimen was cut mechanically, it broke into small pieces, forming the striation shown in figure 18. It is remarkable to see in this figure that all but one of the particles are the same size.

4. Scale Up of the Equipment for Synthesis

The particle size of the product depends largely on the concentration of the vaporised precursor in the plasma gas and there is a strong correlation between production rate and particle size. This has to be taken into account in designing production facilities. At the Forschungszentrum Karlsruhe (Research Centre Karlsruhe) a system for large scale laboratory production was set up. Using this equipment with a reaction tube diameter of 80 mm, particles with mean sizes approximately 8 nm are synthesised with a rate in the range between 50 and 100 g/h. Synthesising particles of sizes below 5 nm reduces the production rate significantly. On the other hand, a remarkably high production rate may be achieved for particles with sizes above 10 nm. This system which is shown in figure 19 is set up with three reaction zones allowing the production of nanoparticles with two different coatings. The precursors may be solid or liquid. Up to three solid precursors are fed with independent gravimetric systems. The feeding rate is from

50 g/h upwards. Alternatively, syringe systems or peristaltic pumps feed liquid precursors to the evaporator. The feeding range of these pumps is sufficiently broad that it is possible to adjust them to the needs of the process. Computer controlling, necessary for industrial production, is possible for all types of feeding systems.

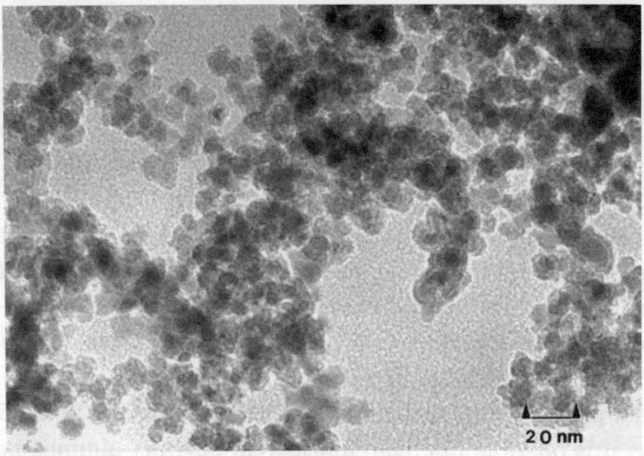

Figure 16a. Maghemite, $\gamma - Fe_2O_3$ particles coated with *PMA*. It is important to note that the particles are nearly uniform in size.

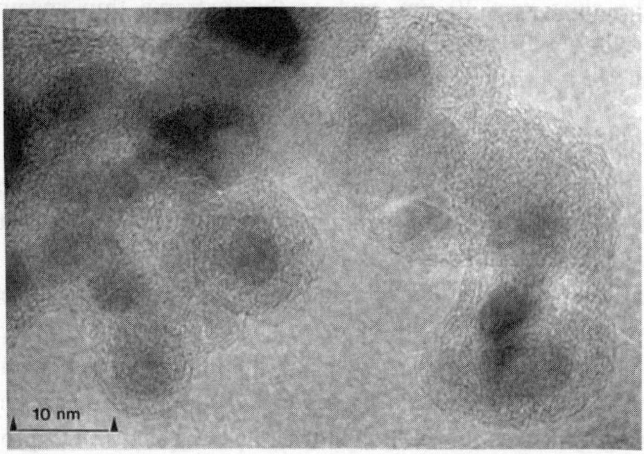

Figure 16b. Maghemite, $\gamma - Fe_2O_3$ particles coated with *PMA* from the same batch as shown in figure 16a. In this figure, it is can be seen that the particles are individually coated. and that the cores do not touch.

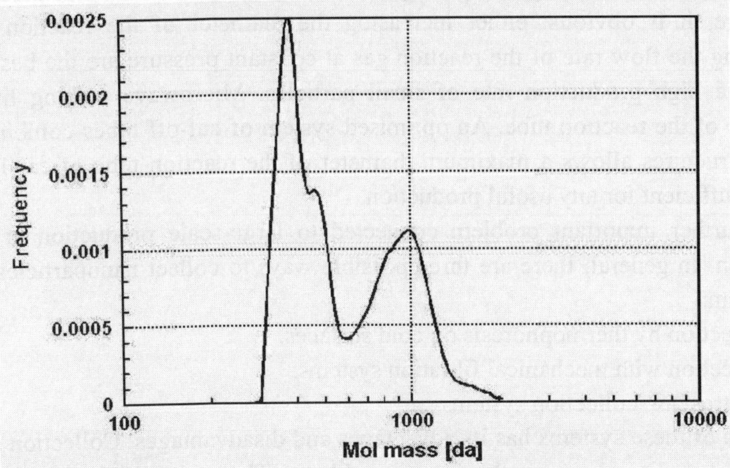

Figure 17. Mol mass distribution of the *PMA* coating of a nanopowder determined by GPC with only very limited polymerisation. The final polymerisation has to be done during hot pressing.

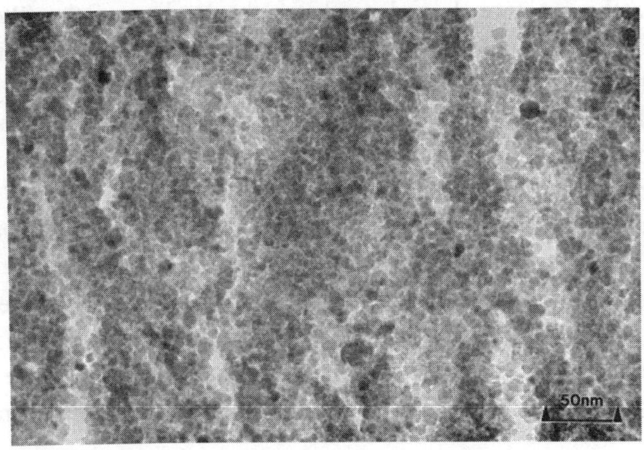

Figure 18. Morphology of a $\gamma - Fe_2O_3 - PMA$ body prepared by hot pressing. The specimen is a 20 nm thick microtome slice. The striation is an artefact from specimen preparation.

The concentration of precursor vapour in the plasma zone controls particle size. Therefore, it is obvious: either increasing the diameter of the reaction tube or increasing the flow rate of the reaction gas at constant pressure are the best way to achieve a high production rate of small particles. Microwave leaking limits the diameter of the reaction tube. An optimised system of cut-off tubes combined with Bragg structures allows a maximum diameter of the reaction tube of ca. 150 mm. This is sufficient for any useful production.

A further important problem connected to large-scale production is particle collection. In general, there are three possible ways to collect nanoparticles from a gas stream:

- collection by thermophoresis on cold surfaces.
- collection with mechanical filtration systems.
- electrostatic collection systems.

Each of these systems has its advantages and disadvantages. Collection systems using thermophoresis cause the fewest problems. The main disadvantage of these simple systems is the poor efficiency. More sophisticated designs using rotating drums work continuously with an efficiency in the range of at least 80%. Mechanical filtration systems are the most efficient: the relatively large pressure drop inherent in them causes severe problems with the microwave plasma process; particularly as the pressure drop depends on the loading of the filters. It is not possible therefore to have constant conditions during synthesis. For large industrial filtration systems, these problems are less severe compared with experimental devices. Electrostatic collection systems have low pressure drop which is independent of filter loading. Continuous cleaning of the system avoids problems with the formation of an electrostatic double layer. Operation of electrostatic collection systems under vacuum is difficult, as the low breakdown voltage in the collector may impede efficient collection.

The system operated at the Forschungszentrum Karlsruhe is centred on the use of microwave technology and each of the three reaction zones needs less than 2 kW of microwave energy. The level of optimisation reached so far reduces the size of the microwave generator to a power level that is not preventing industrial production.

Figure 19. Equipment for continuous production of nanoparticles coated with two layers.

5. Application of Nanomaterials

Nanoparticles, especially coated nanoparticles, produced by the microwave plasma process are highly sophisticated materials and are thus expensive which prevents their use in applications where the gain in performance is small. Only for applications requiring new properties which would be impossible with conventional materials are suitable for these new materials. Coated nanoparticles in particular belong to an entirely new class of material with a high potential for new applications. The combination of two materials with different chemical or physical properties, one acting as core and the other as coating, give rise to the new properties. With increasing understanding of the resulting combined properties, new and interesting applications are possible. Three rather conventional applications for coated nanoparticles are explained in this chapter but further applications based on quantum phenomena are not discussed here.

5.1 Diffusion barrier against grain growth during sintering

A mutual solubility close to nil is a prerequisite for the synthesis of ceramic nanoparticles coated with a second ceramic. The smaller the mutual solubility the better the diffusion barrier provided by the coating which, unlike a the coating material which forms larger grains during sintering, will prevent grain growth of the core particles. This is shown in Figure 15 where the specimen, made from a powder consisting of zirconia coated with alumina, shows the phenomenon clearly. Even after complete sintering, the zirconia particles remain separated in this material, which is essential when physical properties are strongly dependent on the size of the particles, and in addition to this the core particles separated.

5.2 Modification of the chemical surface properties

Nanoparticles, particularly magnetic ones, are excellent carriers of active organic ligands in medicine and pharmacology [37,38] but it is sometimes difficult to attach these ligands to the surface of the nanoparticle of this type, in which case it is important to modify the particle surface by coating with a more suitable for the attachment of a specific organic compound.

5.3 Modification of the physical properties

As mentioned in the introduction, most of the special properties characteristic of nanomaterials are the properties of a single isolated particle. By reducing the interaction of the particles with a distance holder, it is possible to produce technical modifications while maintaining the "nanospecific" properties. A typical example of this is the superparamagnetism, a property which is strongly dependent on grain size. Superparamagnetism is explained in detail below to demonstrate the possibilities and problems of the systems.

Particles of maghemite, $\gamma - Fe_2O_3$, a ferrimagnetic material with sizes below ca. 8 nm, are superparamagnetic. [39] Superparamagnetism is limited to isolated single domain particles, characterised by a thermally fluctuating vector of magnetisation. A particle fulfilling the condition

$$kT > Kv \tag{7}$$

(K ... constant of magnetic unisotropy, v ... volume of the particle; Kv ... energy of magnetic unisotropy, kT ... thermal energy) is superparamagnetic [39,40] and the coercive ability of such particles is zero at temperatures above:

$$T_B = \frac{Kv}{k} \tag{8}$$

The temperature T_B is the blocking temperature. The thermal relaxation time τ, limits the maximum frequency of application. According to Néel [40]

$$\tau = \tau_0 \exp(\frac{Kv}{kT}) \tag{9}$$

describes the relaxation time, and τ_0 is in the range between 10^{-9} and 10^{-13} sec., therefore, at room temperature the frequency of thermal relaxation may be well beyond 1 GHz. Formulae (8) and (9) clearly show that the blocking temperature and, more importantly the relaxation time, are volume-dependent. This demonstrates the importance of an extremely narrow particle size distribution.

After sintering such a powder, the product no longer exhibits any magnetism. Why is this? Grain growth occurred during sintering when the grains get larger than ca. 25 nm $\gamma - Fe_2O_3$ and transform to the hexagonal $\alpha -$ phase which is antiferromagnetic. If one assumes no grain growth during sintering, the final product would no longer exhibit superparamagnetism because of the magnetic dipole-dipole interaction in which case the originally superparamagnetic particles form larger magnetic clusters, which are no longer superparamagnetic. Coating $\gamma - Fe_2O_3$ particles with a second phase, a ceramic or a polymer phase, avoids these problems. Figure 16 and 18 show, for example, $\gamma - Fe_2O_3$ particles coated with *PMMA*. Figure 20 depicts the magnetisation curve of this material. As is characteristic for a superparamagnetic material, the magnetisation curve is free of any hysteresis.

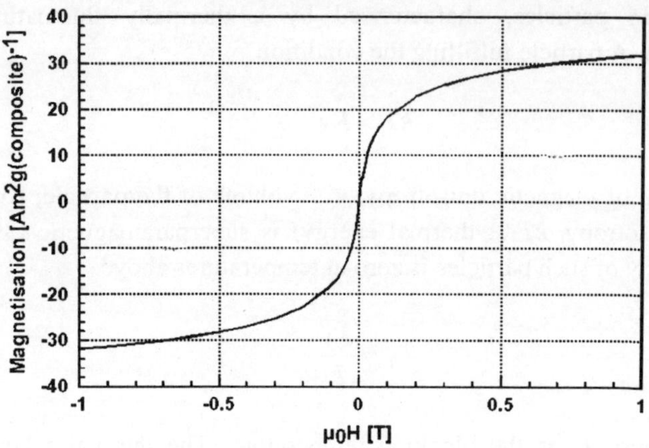

Figure 20. Magnetisation curve of *PMMA* coated $\gamma - Fe_2O_3$.

The Langevin function *L(m,H,T)* describes the magnetisation *M* of superparamagnetic materials consisting of *n* particles with a magnetic moment *m* in the field *H* at the temperature *T* with sufficiently good approximation.

$$M = nmL(m,H,T) = nm(cth(\frac{mH}{kT}) - \frac{kT}{mH}) \tag{10}$$

In reality, a specimen consists of particles of different size. Assuming *I* classes each with the moment m_i and the number n_i of particles in it, the magnetisation curve can be described by:

$$M = \sum_{i=1}^{I} n_i m_i L(m_i, H, T) \tag{11}$$

By fitting the values $n_i(m_i)$ to the magnetisation curve, it is possible to calculate the distribution of magnetic moments in a specimen. Figure 21 depicts the distribution of magnetic moments derived from the magnetisation curve in Figure 20.

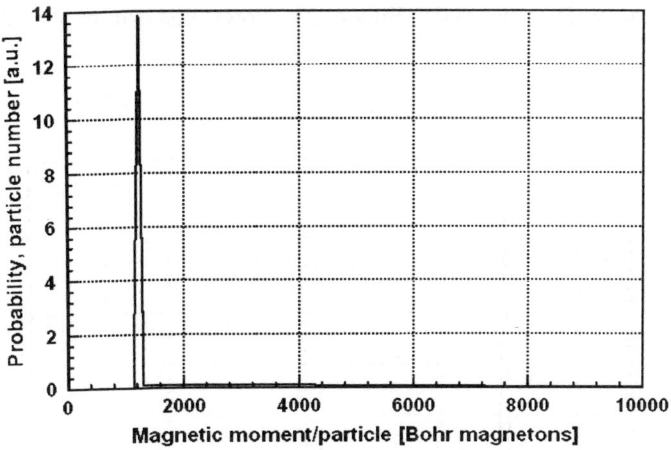

Figure 21. Distribution of magnetic moments of *PMMA* coated $\gamma - Fe_2O_3$ derived from the magnetisation curve shown in Figure 20.

The magnetisation is proportional to the volume of the particles and thus $(m_i)^{0.33}$ is proportional to the particle size. This leads to the size distribution shown in Figure 22. The absolute size in the nanometer is obtained by calibration of the distribution with micrographs. The important point in Figures 21 and 22 is that the distribution of the particle size for *PMMA* coated $\gamma - Fe_2O_3$ is extremely narrow. This is necessary for any application of these materials, because the blocking temperature is linearly dependent on the particle volume and the relaxation time increases exponentially with particle size. A broad distribution of particle size thus leads to a material with poorly defined properties which makes any application of these properties impossible.

The superparamagnetism shown so far has relaxation times of less than 100 s, the time taken to measure one point in the magnetisation curve. For technical applications, shorter relaxation times in the range of less than 1 ns are necessary and therefore the ultimate indication of superparamagnetism is obtained by Mößbauer spectroscopy where testing relaxation frequencies are shorter than the Lamor frequency of the iron nucleus. In this case the nuclei are "unaware" that there is a high magnetic crystal field which leads to Mößbauer spectra showing only the quadruple doublet of nonmagnetic material instead of the sextet which is normally observed for magnetic materials. Figure 23 shows this for a $\gamma - Fe_2O_3$ - PMA composite specimen.

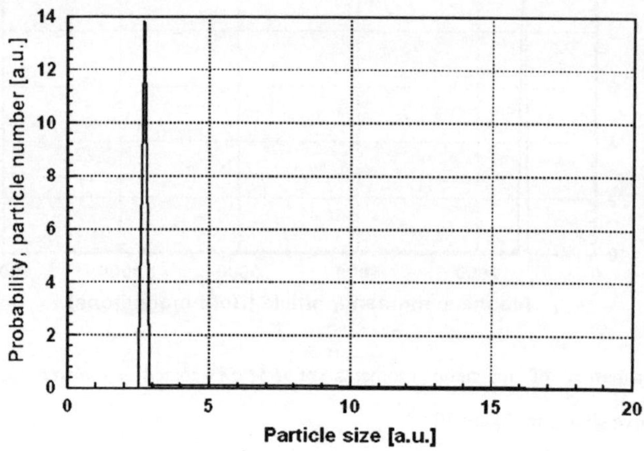

Figure 22. Particle size distribution of *PMMA* coated $\gamma - Fe_2O_3$ derived from the magnetisation curve depicted in Figure 20.

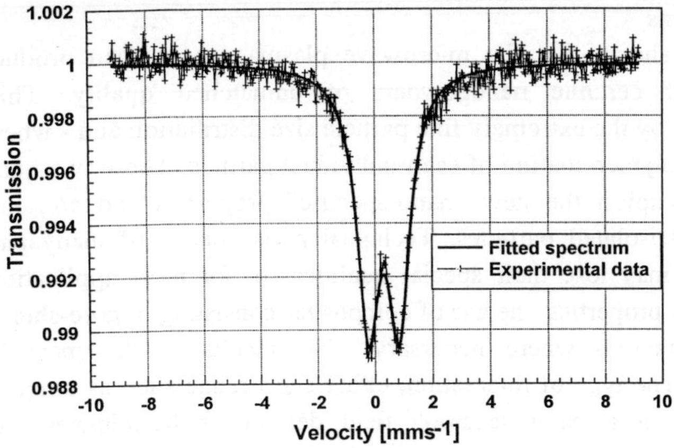

Figure 23. Mößbauer spectrum of a $\gamma - Fe_2O_3 - PMA$ composite specimen.

One of the properties required for technical application is susceptibility in a frequency range beyond the range of classic ferrites. Compared with classic ferrites, superparamagnetic ferrites have the advantage for application of lower losses and higher maximum frequencies. The resonance frequencies of these materials are well beyond 2 GHz. The data for the complex susceptibility of a $MnFe_2O_4 - PMMA$ ferrite are depicted in figure 24.

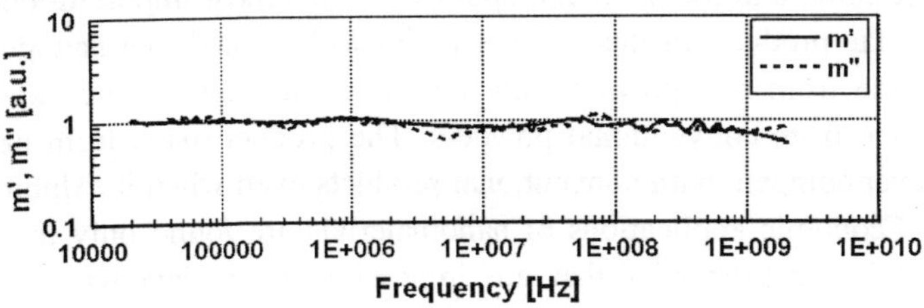

Figure 24. Real and predicted susceptibility for a nanocomposite containing $MnFe_2O_4$ - *PMMA*.

In Figure 24, the part m' decreases slightly with frequency. The influence of the electron spin resonance was not observed in the frequency range up to 2 GHz. Which should be compared with the maximum frequency for classical ferrites of ca. 200 MHz.

6. Conclusions

It has been shown that the microwave plasma process can produce significant quantities of ceramic nanopowders of unmatched quality. This quality is characterised by the extremely fine particle size distribution and - where necessary - ceramic or polymer coating of each individual particle. These two qualities make it possible to exploit the new "nano-specific" properties and are, in most cases, properties of isolated particles. Technical parts consist of many particles which interact and may lose their special qualities so, for most applications exploiting specific nano-properties, the use of composites consisting of core-shell particles will prevent interaction where necessary. This results in the most homogeneous composites. The type of interaction, either short-range such as tunnelling, or long-range such as in a magnetic dipole field, determines the thickness of the coating. Using the microwave plasma process, the synthesis of coated particles is possible because the particles leave the plasma zone with electric charges of equal sign. This prevents the particles from forming a cluster and also limits particle growth which would lead to extremely fine particle size distribution. The microwave plasma process works continuously: a complete system consists of automatic gravimetric dosing systems and a continuous particle precipitator. Based on such components, a demonstration plant was set up which was able to produce 100 g batches of nanoparticulate oxide, nitride, sulphide, or selenide powders, coated or uncoated, and with mean particle sizes in the range from 5 to 20 nm.

The last point, and for industrial applications the most important one, concerns the price of the product. In this context it should be noted that nanopowders, and products made from nanopowders, are always significantly more expensive than products made from conventional powders. The product made from nanopowders can thus never compete with conventional products even when it exhibits improved properties. Economic applications of nanomaterials are found only in areas where the demand is for properties that are impossible to achieve with conventional materials. The crucial point is to search for new rather than for improved properties. If new properties can be used to solve a particular problem, then the price is no longer so important: in applications such as this, the products produced by the microwave plasma process are unmatched with respect to particle size distribution and perfect coating.

References

1. Gleiter H., Nanocrystalline materials. *Prog. in Mat. Science* **33** (1989) 223-315.
2. Vollath D. & Szabó D. V., Synthesis of ceramic nanocomposites in a microwave plasma. In *Advances in Science and Technology, Vol. 3B Ceramics: Charting the Future, ed. by P. Vincencini,* Florenz 1995, pp. 1261-1268.
3. Vollath D. & Szabó D. V., Nanocoated particles: a special type of ceramic powder. *Nanostructured Materials* **4** (1994) 927-938.
4. Karch J., Birringer R. & Gleiter H., Ceramics ductile at low temperature. *Nature* **330** (1987) 556-558.
5. Gleiter H., Materials with ultrafine microstructures: retrospectives and perspectives. *Nanostructured Materials* **1** (1992) 1-19.
6. Hahn H., Eastman J. A. & Siegel R. W., Processing of nanophase ceramics. *Ceramic Transactions* **B1** (1988) 1115-1122.
7. Günther B. & Kumpmann A., Ultrafine oxide powders prepared by inert gas evaporation. *Nanostructured Materials* **1** (1992) 27-30.
8. LaiHing K., Cheng P. Y. & Ducan M. A., Ultraviolet photolysis in a laser vaporization cluster source: synthesis of novel mixed-metal clusters. *J. Phys. Chem.* **91** (1987) 6521-6525.
9. Kriechbaum G. W. & Kleinschmit P., Superfine oxide powders-flame hydrolysis and hydrothermal synthesis. *Angew. Chem. Adv. Mater.* **101** (1989) 1446-1453.
10. Vollath D., Sickafus K. & Varma R., Synthesis of nanocrystalline powders for oxide ceramics by microwave plasma pyrolysis. In *Mat. Res. Soc. Symp. Proc.* **269** (1992) 379-384.
11. Vollath D. & Sickafus K. E., Synthesis of nanosized ceramic oxide powders by microwave plasma reactions. *Nanostructured Materials* **1** (1992) 427-437.
12. Vollath D. & Sickafus K. E., Synthesis of ceramic oxide powders in a microwave plasma device. *J. Mater. Res.* **8** (1993) 2978-2984.
13. Vollath D. & Sickafus K. E., Synthesis of nanosized ceramic nitride powders by microwave supported plasma reactions. *Nanostructured Materials* **2** (1993) 451-456.

14. Chang W., Skandan G., Hahn H.,Danforth S. C. & Kear B. H., Chemical Vapor condendation of nanostructured ceramic powders. *Nanostructured Materials* **2** (1994) 345-351.
15. McDonald A. D, *Microwave Breakdown in Gases*. John Wiley & Sons, New York, 1966.
16. Vollath D., Synthesis of nanocrystalline ceramic powders in a microwave plasma (in german). *KfK-Nachrichten* **25** (1993) 139-144.
17. Vollath D., A cascaded microwave plasma source for synthesis of ceramic nanocomposite powders. In *Mat. Res. Soc. Symp. Proc.* **347** (1994) 629-634.
18. Vollath D., Szabó D. V. & Seith B., German Patent 196 38 601.2-43 (1997).
19. Vollath D., Szabó D. V. & Fuchs J., Synthesis and properties of ceramic – polymer composites. *Nanostructured Materials* **12** (1993) 433-438.
20. Pawlow P., Upon the dependence of the melting point from surface energy of a solid (in german). *Zeitschr. f. physik. Chemie* **65** (1908) 1-35.
21. Hill T. L., Termodynamics of small systems, edited by W. A. Benjamin, Inc., New York (1993) pp. 89-132.
22. Forker M., Schmidberger J., Szabó D. V. & Vollath D., Perturbed-angular-correlation study of phase transformations in nanoscaled Al_2O_3-coated and noncoated ZrO_2 particles synthesized in a microwave plasma. *Phys. Rev. B* **61** (2000) 1014-1025.
23. Roth P, private communication (2000).
24. Hospital A. & Roth P., *In-situ* Mass Growth Measurements of Charged Soot Particles from Low Pressure Flames. *23rd Symposium (International) on Combustion*, The Combustion Institute, Pittsburgh, PA, pp. 1573 – 1579, 1990.
25. Roth P. & Hospital A., Mass Growth and Coagulation of Soot Particles in Low Pressure Flames. *24th Symposium (International) on Combustion*, The Combustion Institute, Pittsburgh, PA, pp. 981 – 989, 1992.
26. Roth P. & Hospital A., design and test of a Particle Mass Spectrometer (PMS). *J. Aerosol Science* **25**, 1, (1994) 61-73.
27. Lindackers D., Strecker M. G. D., Roth P., Janzen C., & Pratsinis S. E., Formation and growth of SiO_2 particles in low pressure $H_2/O_2/Ar$ flames doped with SiH_4. *Combust. Sci. and Tech.* **123** (1997) 287-315.

28. Vollath D. Szabó D. V. & Haußelt J., Synthesis and properties of ceramic nanoparticles and nanocomposites. *J. Europ. Ceram. Soc.* **17** (1997) 1317-1324.
29. Tenne R. Margulis L. Genut M & Hodes G., Polyhedral and cylindrical structures of tungsten disulphide. *Nature* **360** (1992) 444-446
30. Margulis L., Saltra G., Tenne R. & Tallanker M., Nested fullerene-like structures. *Nature* **365** (1993) 113-114.
31. Feldmann Y., Wasserman E., Srolovitz D. J. &Tenne R., High-rate, gas-phase growth of MoS_2 nested inorganic fullerenes and nanotubes. *Science* **267** (1995) 222-225.
32. Vollath D. & Szabó D. V., Synthesis of nanocrystalline MoS_2 and WS_2 in a microwave plasma. *Materials Letters* **35** (1998) 236-244.
33. Vollath D. & Szabó D. V., Nanoparticles from compound with layered structures. *Acta mater.* **48** (2000) 953-967.
34. The American Ceramic Society, *Phase Diagrams for Ceramists*, Westerville, OH, USA (1975), 135.
35. Szabó D. V. & Müllejans H., Characterization of nanocomposites by electron microscopy methods. In *Proceedings of the ICEM 13*, **Vol. 2B** (1994) 379-380, Les Éditions de Physique, Paris.
36. J. W. Matthews, Misfit Dislocations, in: *Misfit Dislocations in Solids*; ed.: F. R. N. Nabarro, North Holland, Amsterdam, 1979.
37. Pfefferer D., Wagner S., Kresse M., Taupitz M. & Ebert W., MR-Aniography with Superparamagnetic Iron Oxide Particles: Optimisation of Particle Size and Physico-chemical Properties. *ECR '95, Wien (A),05.-10.03.95*
38. Jordan A., Wust P., Fähling H., Jahn W., Hinz A. & Felix R., Inductive heating of ferrimagnetic particles and magnetic fluids: physical evaluation of their potential for hyperthermia. *Int. J. Hyperthermia* **9** (1993) 51-68.
39. Vollath D., Szabó D. V., Taylor R. D., Willis J. O. & Sickafus K. E., Synthesis and properties of nanocrystalline superparamagnetic γ-Fe_2O_3. *Nanostructured Materials* **6** (1995) 941-944.
40. Néel L., *Comt. Rend.* **228** (1949) 664.

CHAPTER NINE

DEPOSITION OF ELECTRON BEAM EVAPORANT IN A LOW VACUUM GAS FLOW ENVIRONMENT

J.F. Groves and H.N.G. Wadley

University of Virginia, Thornton Hall A127
Materials Science and Engineering Department
Charlottesville, Virginia, 22904-4240, USA
jgroves@virginia.edu

1. Motivating Development of a New Vapor Deposition Technology

Since Grove observed metal deposits sputtered from a glow discharge in 1852 [1] and Faraday experimented with metal wire explosion in an inert atmosphere five years later [2], researchers have synthesised a multitude of materials from the vapour phase, sought to understand the properties of those materials, and explored their scientific and technological applications. Over time, engineers have realised that vapour-phase-created materials (i.e. materials created via physical vapor deposition (PVD) or chemical vapor deposition (CVD)) can meet the design requirements of countless products. In some cases a thin film meets the engineering need (e.g. an environmental barrier [3] or wear resistant [4] coating). Sometimes vapour phase processing represents the only method by which the desired material can be applied (e.g. optical gold coatings for lenses and mirrors [3] or aluminum metallisation layers and copper seed layers for semiconductor interconnections [5]). In other instances vapour phase synthesised films have properties and atomic structures which are not found in bulk materials (e.g. giant magnetoresistive (GMR) multilayers [6], thermal barrier coatings (TBC) [7], and functionally graded materials (FGM) [8]). Finally, thick and thin films play important roles in numerous other applications as well such as tool coatings, catalysis systems, coatings for biomedical devices, and sensor system multilayers.

Since the 1940's researchers have developed vapour fabrication methods to synthesise thin film materials (e.g. electron beam evaporation [9], diode and magnetron sputtering [10], CVD [10], reactive evaporation (RE) and activated reactive evaporation (ARE) [11, 12], ion plating [13], and ion-beam assisted deposition [14]). Even as these and other technologies are increasingly employed in manufacturing systems, new materials demands stretch the scientific limits of vapour phase processing capability. Indeed, despite the sophistication of today's vapour processing tools, there appear to be variations of these techniques which could be developed to fabricate the products of today and tomorrow that demand higher performance, performance that can be created only via precise, highly sophisticated materials processing technologies.

The materials properties of a vapour-deposited thick or thin film can be affected by each of the five steps in the vapour phase synthesis process: vapour *creation*, vapour *transport*, vapour *adsorption* on to the substrate, adatom *diffusion across the* substrate *surface*, and atom movement by *bulk diffusion* through the growing film lattice to a final position. Ideally a vapour deposition technique would possess an ability to manipulate each of these steps for optimal film synthesis. In reality, the method of precise manipulation is often poorly understood and difficult to incorporate into an actual processing tool.

This chapter describes the development of a new physical vapour deposition technology which seeks to impart material processing flexibility to the process steps listed above through a unique assembly of technology which deposits electron beam (e-beam) evaporant in a low-vacuum gas flow environment. The chapter will show how understanding of the technology has been generated via experimental and modelling observations of vapour transport and deposition. Selected results from both types of investigation will be presented. Experimental study has included optical observations of vapour location during transport from source to deposition surface, and deposition efficiency studies of coated substrate surfaces. Modelling observations of the vapour transport and deposition characteristics of the technology have been made using Direct Simulation Monte Carlo (DSMC) and bimolecular collision theory techniques. In general, study of this material deposition method should contribute to our understanding of how atoms can be intelligently manipulated during thick and thin film creation for the synthesis of highly-engineered coatings. As reported in this Chapter, these results are already being used to guide further development of the technology and to identify applications which might benefit from the particular materials processing characteristics of the technology.

2. Depositing Electron Beam Evaporant from a Gas Flow

The many applications listed at the outset of this article demand a myriad of performance characteristics from their coating layers. To meet these diverse needs, the various film layers must often be formed with distinctly different elemental compositions and internal atomic structures. While pure single element coatings possess the best properties for some applications, precise metal alloys or multielement ceramic systems are frequently required for others. Similarly, the coating structure often has a significant bearing upon final system performance. For a given application, the best film properties might be offered by an amorphous, quasicrystalline, or nanocrystalline material rather than a porous, single crystal, or fully dense polycrystalline layer (or multilayer stack).

Clearly today's applications require a broad range of film compositions and atomic structures for optimum performance. The ability of a particular vapor deposition process to deliver these compositions and structures is highly dependent upon whether or not the process can affect each of the five steps of vapour phase material synthesis (creation, transport, adsorption, surface diffusion, and bulk diffusion). In an attempt to create such a tool, a vapour deposition system has recently been developed at the University of Virginia which performs e-beam evaporation in a low vacuum, flowing gas stream environment. This technology seeks to harness the best features of e-beam material creation, to introduce those characteristics into a medium and low vacuum (1 microbar - 10 millibar) flowing gas vapour transport environment, and to create high performance coatings by controlled vapour atom adsorption, surface diffusion, and bulk diffusion.

2.1. Vapor creation using an electron beam

Use of e-beams to create material vapour streams for coatings has been widespread since the 1960's, partly because of the ability of e-beams to create vapour from a wide variety of materials rapidly and cleanly. High vacuum e-beam systems (e.g. operating at 10^{-3} - 10^{-1} microbar) have been used to vaporise and then deposit not only pure elements like aluminum, zinc, gold, and silver which are relatively easy to process, but also more difficult elements with low vapour pressure such as molybdenum, tungsten, and carbon, and highly reactive elements such as niobium, titanium, and tantalum [15]. In addition to an ability to vaporise pure elements, e-beam systems have shown a reasonable ability to create material from metal alloy and ceramic sources. E-beam systems evaporate and deposit all of these elements by bringing the heat source (electrons) cleanly into direct contact with the source material, often contained as a "skull" melt inside a water-cooled crucible (Fig. 1a). A crucible is frequently used to contain the source material because it maintains solid source material (the "skull") between the crucible wall and the molten evaporant pool, preventing vapour source contamination from the crucible.

Although widely successful, e-beam evaporation of alloys and ceramics often proves to be challenging. For metal alloy systems with a vapour pressure ratio as high as 1000:1 between alloy elements, stoichiometrically correct evaporation and deposition from a single crucible source is possible [16, 17]. For alloys with elements having a greater vapour pressure ratio, lower melting point (and higher activity) material often needs to be wire-fed into a pool of the more refractory material [18], or in some cases separate element evaporation from adjacent crucible sources is necessary [9, 15]. Evaporation of elemental compounds can be especially challenging in e-beam systems. For compounds which are poor conductors of electricity and heat, some combination of reduced e-beam power densities (below $2x10^7$ W/m^2) [1], specialised e-beam scan patterns [1], and partially dense source materials (e.g. 60% dense yttria-stabilized zirconia for TBC applications [36]) are usually necessary to prevent source cracking and to ensure controlled vapour stream generation. Scientists and engineers have also been presented with problems during the evaporation of ceramics by incomplete evaporation of material at the edge of the crucible which has necessitated the use of special e-beam scan patterns for the control of these spikes of solid ceramic protruding from the crucible edge. Furthermore, ceramic evaporation often requires the introduction of oxygen into the process chamber, to compensate for oxygen lost from the compound during the evaporation process.

Although similar vapour pressure sources, wire-feeds, adjacent crucibles, special scan patterns, and partial pressures of reactive species have allowed engineers to vaporise and deposit stoichiometrically correct layers in high vacuum, e-beam processing of such sophisticated, compositionally-correct alloy and ceramic systems is often tremendously difficult. Vaporisation from alloy sources with different vapor pressures can occur only after the system has reached evaporation equilibrium [9]. Deposition using wire feeds can create compositional differences across the melt pool as the wire material diffuses into the main source. Alloy processing from multiple crucibles in high vacuum creates a stoichiometrically correct deposit only in that region above the crucibles where the vapour clouds of the neighboring crucibles intersect [9]. Introduction of too much oxygen into the system during ceramic processing can lead to significant numbers of gas/vapour atom collisions which decrease vapour atom energy during transport. This decrease in energy can affect the atomic structure of the growing film in an adverse way [11]. It is clear that, in common with many of today's vapour phase processing technologies, standard e-beam systems have their positive features and their challenging limitations.

2.2. Vapor transport in a flowing gas stream

After atomistic vapour has been created with an e-beam, vapour transport to the substrate can occur either as a result of the vapour creation process itself (i.e. evaporation atoms can be propelled through the process chamber by means of their thermal evaporation energy), or as the result of external factors acting upon the individual vapour atoms (i.e. interaction with electric or magnetic fields or background gas atoms in the chamber can modify atom velocity vectors). Indeed, the particular PVD/CVD process and conditions used to generate a vapour stream have been shown to influence the way in which a film grows on a surface. The critical factors affecting growth are the spatial distribution, the angle of incidence, kinetic energy, deposition efficiency, and form (e.g. monatomic or multiatom clusters) of vapour atoms reaching a substrate. Thornton et al were the first to begin mapping out these relationships experimentally [30, 53, 54]. More recently Zhou, Johnson, and Wadley have provided significant new insights through molecular dynamics modelling [55].

Often, the inherent characteristics of a high vacuum e-beam vapour stream (i.e. the deposition efficiency and spatial, angular, and energy distributions) do not coincide with the necessary vapour stream characteristics for a specific application. Researchers have sought to modify vapour distributions. Raising the pressure above 10^{-2} millibars in the background processing chamber has produced a more uniform vapour stream and enhanced non-line-of-site coatings [19, 20]. Nichols, Rossnagel and others [21] have successfully used electrostatic collimation in sputtering systems to manipulate the ionized vapour flux angular distribution. In e-beam systems, Mattox [13] and others [24] have found that interaction of the e-beam with the gas in the chamber can create an ionized plasma of vapor atoms and chamber gas atoms. By applying a negative electrical bias (~ 100 V) to the deposition substrate, both ionised gas and vapour atoms can be accelerated towards the substrate as part of an "ion plating" process, imparting kinetic energies of 10-10,000 eV to depositing atoms [13]. These elevated kinetic transport energies influence the subsequent adsorption and diffusion steps of the film synthesis process.

During the 1990's, materials processing efforts by Eastman, Halpern, and others [25, 26] have increased understanding of vapour transport during materials processing, this time in a low or medium vacuum, flowing gas environment. Eastman et al used e-beam evaporation to create nanophase γ-Al_2O_3 clusters with a mean grain size of 2.5 nm in a 1 millibar oxyge- rich environment, while Halpern et al. deposited resistively-evaporated gold by transporting vapour to a substrate in a helium gas jet at pressures of approximately 1 millibar. More recently Halpern and others have deposited polymers and ceramics in low vacuum environments [22, 23].

The use a flowing gas jet to transport thermally evaporated material to a substrate for efficient deposition has been reported by numerous researchers [26, 27, 28, 30] (Fig. 1b). Reported results from molecular beam separators and the Jet Vapor Deposition™ process have suggested that these techniques can affect vapour transport (i.e. spatial, angular, and energy distributions), beneficially modifying deposition efficiencies, deposit compositions, and deposit forms (single atom, compound, or cluster). Although the exact influence of the flowing gas upon the vapour stream has not been clearly defined in the literature, the general concept has several appealing characteristics which, when combined with the known advantages of e-beam vaporisation, appear to present an opportunity to create a new, useful and far from obvious technique for synthesising film layers with precise composition and atomic structure characteristics.

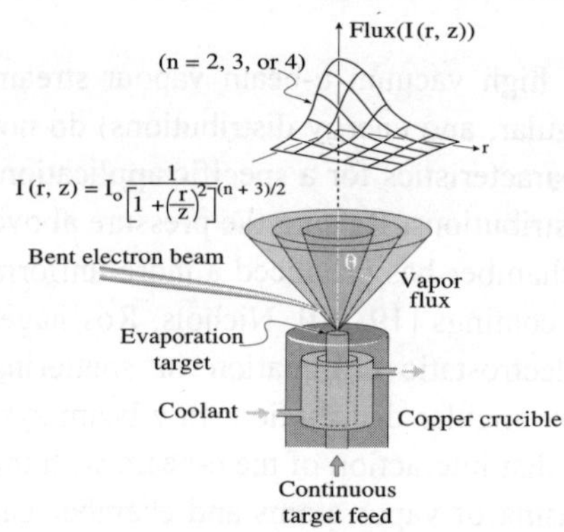

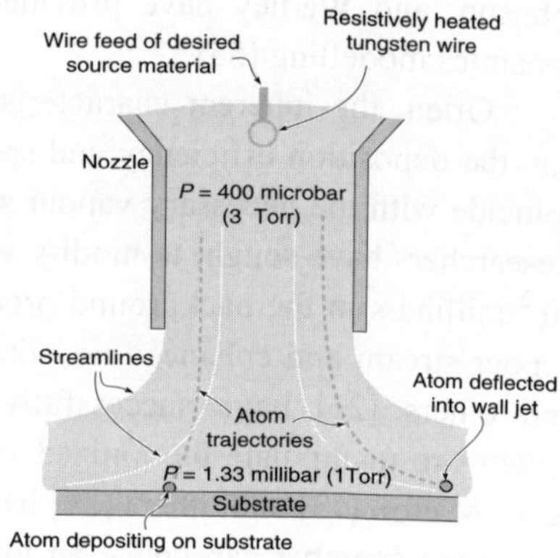

Figure 1 a) Conventional e-beam evaporation creates a diverging vapour stream in a high vacuum. b) Molecular beam separators and similar processes use a nozzle and gas flow to focus vapour on to a substrate.

2.3. *Directed vapor deposition*

Based on a knowledge of state-of-the-art e-beam vaporization and low vacuum vapour transport, a low and medium vacuum e-beam evaporation system (1 microbar - 10 millibar) has been created. This new concept of directed vapour deposition (DVD) is based upon the known and theoretical capabilities of e-beam vaporization and carrier gas stream vapour transport. By combining these technological solutions, it has been posited that a new deposition method could be constructed with enhanced flexibility for manipulation of the five stages of vapour deposition - vapour creation, transport, adsorption, surface diffusion, and bulk diffusion. In a departure from conventional high vacuum (10^{-3} - 10^{-1} microbar) e-beam vaporization methods [2, 9, 15], DVD has been designed to function in a vacuum environment where a carrier gas stream incorporates evaporants for transport and deposition (Fig. 2) [31]. The studies in this chapter are designed to evaluate the capability of DVD's in improving upon the current position.

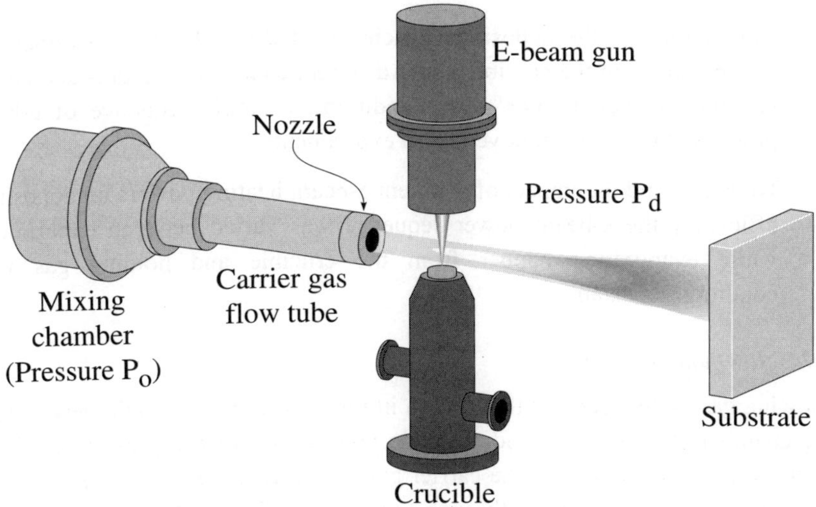

Figure 2 In a directed vapour deposition system, e-beam evaporant from the crucible is transported to the substrate in a carrier gas stream.

3. Experimental Study of "Directed Vapour Deposition"

To improve understanding of all five stages of vapour deposition in DVD, several sets of experiments have been performed. Fig. 2 shows the general configuration of the DVD system used for the experiments described here.

1. Determine the gas stream's structure without the influence of a vapour stream. The e-beam was propagated through an argon carrier gas flow while the crucible was empty, and this resulted in luminescence of the gas stream that facilitated visual study of the stream's structure.

2. Studies of the interaction of the gas and vapour streams were performed under different processing conditions. The e-beam was used to evaporate copper from a water-cooled crucible and the evaporated material was entrained in either an argon or helium gas flow for transport towards the substrate. Throughout the experiments, copper evaporation generated a green luminescence which provided a convenient means of visualizing transport: this allowed the various processing parameters to be modified and the relationship between process parameters and vapour atom trajectories to be studied.
3. Examination of the deposition efficiency of the technology. Copper was evaporated from the crucible under different chamber pressures and carrier gas flow velocities. For these conditions, a single sequence of e-beam power settings was employed in all experiments.
4. To understand the impact of different e-beam heating powers on deposition efficiency the e-beam power sequence was varied between experiments while evaporating copper from the crucible and holding gas flow conditions constant.

3.1. *Gas flow fundamentals*

In a flowing gas system such as that shown in Fig. 2, the pressure ratio between the mixing chamber at pressure P_o and the main process chamber at pressure P_d determines the velocity flow field of the carrier gas as it leaves the nozzle. Estimates of the maximum velocity attainable through use of a gas flow and nozzle technology such as that of DVD can be determined using one-dimensional equations for isentropic flow of a compressible fluid [32]. Once the velocity of the flow field can be verified, a simple calculation can be performed to determine the maximum kinetic energy that can be given to a vapour atom entrained in that flow.

The important governing relationships between pressure, temperature, Mach number, and gas stream velocity are given by:

$$\frac{P_o}{P_d} = \left[1 + \frac{\gamma - 1}{2} M^2\right]^{\gamma/(\gamma - 1)} \tag{1}$$

and

$$U = M\sqrt{\gamma R_s T} \tag{2}$$

where P_o is the upstream pressure before the nozzle (millibar), P_d is the downstream pressure at the nozzle or in the chamber (millibar), γ is the ratio of specific heats (5/3 for helium and argon), M is the flow's Mach number, U is the carrier gas stream speed (m/sec), T is the absolute temperature (K), and R_s is the specific gas constant (2077 J/(kg K) for helium, 208.1 J/(kg K) for argon). Early measurement of P_o and P_d in the DVD system indicated that the carrier gas stream flow in the system should be reaching supersonic velocity based on equation (1) [29, 30, 31]. As a result, the expansion of the carrier gas out of the mixing chamber and flow tube should be generating a gas atom spatial and velocity vector distribution typical of supersonic nozzle expansions.

In the literature, Scoles [33] indicates that the *distance* of the Mach disk[1] from the end of the flow tube and nozzle can be calculated with reasonable accuracy given the nozzle diameter and pressure ratio (Mach number) used in the system:

$$x_m = 0.67d\left(\frac{P_o}{P_d}\right)^{1/2} \tag{3}$$

where x_m is the Mach disk distance from the nozzle exit (m), and d is the nozzle diameter (m). The literature also indicates that the *diameters* of the zone of silence[2] and of the Mach disk should be ~0.75 x_m and ~0.50 x_m, respectively [33]. These dimensions are not as well defined as the Mach disk location, being more sensitive functions of the specific heat ratio γ and of P_o/P_d [33].) Adamson and Nicholls [34] have explained that lower Mach numbers decrease the diameter of the zone of silence and the Mach disk.

Given this fundamental understanding of how the gas flow in the DVD system should behave, experiments were undertaken to confirm system performance so that estimates could be made of vapor atom kinetic energy in the system. Studies were also performed to see how introduction of a vapour flux into the carrier gas stream affected gas and vapour flow through the process chamber.

---

[1]Mach disk - zone in which atoms that have left the nozzle abruptly slow down for the first time to subsonic velocities.

[2]zone of silence - initial supersonic flow region just outside the nozzle.

3.2. Visualizing the carrier gas flow

For the first set of experiments, the e-beam gun was turned on at 1.2 - 1.8 kW and passed through argon flows of 10 - 20 standard liters per minutes (slm)[3]. Based on equation (1), pressure ratios (P_o/P_d) corresponding to Mach 1.7 (1.27 cm diameter nozzle) and Mach 2.5 (0.85 cm diameter nozzle) conditions were established in the system, and the resulting flow structure was recorded [30]. For each set of conditions explored, the major flow structures of a supersonic free jet expansion (i.e. the zone of silence, the Mach disk, and the secondary "barrel" shocks [30, 34]) were observed (Fig. 3). While the zone of silence is just visible in the vicinity of the e-beam for the Mach 1.7 flow experiments, it is larger and more clearly visible in the Mach 2.5 flow runs as predicted in the literature. The sharp transition of the Mach disk is also dramatically evident. Estimates of the experimental separation between nozzle and Mach disk were made, which indicated a distance of 2.0 cm for the Mach 1.7 flows and 3.2 cm for the Mach 2.5 flows. These distances are within 10-15% of those predicted by equation (3). The carrier gas velocity in the secondary barrel shocks is expected to be supersonic, but the exact velocities are unknown.

These visualization results and previously reported gas flow characteristics of DVD [29, 30], strongly suggest that the carrier gas flows of this process achieve supersonic velocities. To determine the effect of these velocities upon vapour atom kinetic energies, equations (1) and (2) can be combined with the basic physics equation for kinetic energy and the following equation to determine the maximum energy of individual vapour atoms in the flow.

$$\frac{T_o}{T_d} = 1 + \frac{\gamma - 1}{2} M^2 \tag{4}$$

In this equation T_o is the upstream temperature (K) and T_d is the downstream temperature (K). Given a Mach number of 2.5 and an upstream gas temperature of 300 K, a copper atom will reach a maximum kinetic energy of 0.7 eV in helium and just 0.07 eV in argon, assuming that the vapour atom is entrained in the main, supersonic flow. This compares with an initial, vaporisation-induced energy of about 0.2 eV [30]. Thus, these first order calculations suggest that DVD can modify

---

[3]standard litre - one litre of gas at atmospheric pressure, room temperature.

the initial energy of the vapour atoms through use of the carrier gas stream: however, it is not certain from these results that the vapour atoms can carry these energies to the deposition surface.

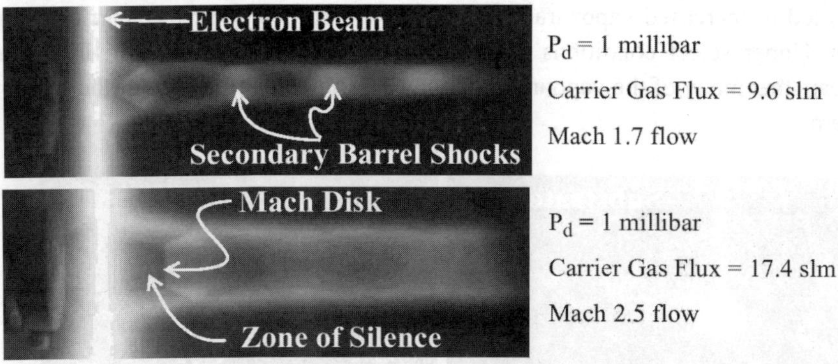

Figure 3 DVD argon gas flow structure for different Mach number flows.

3.3. Visualising the interaction between carrier gas and vapour flows

To understand more about the relationship between carrier gas flow and vapour atom transport, photographs were taken of carrier gas/vapour stream /substrate interaction. Conditions that were varied included carrier gas flux (slm), carrier gas Mach number (equation (1)), carrier gas type (argon vs. helium), and e-beam power (kW). While a selection of results will be described here, the complete results are presented elsewhere [30].

The first vapour visualization experiments explored the effect of varying the carrier gas flux upon vapour transport. While holding the carrier gas stream Mach number approximately constant, helium carrier gas stream fluxes were varied from 0.5 slm to 5.0 slm and then to 50 slm. While these gas flow conditions were transmitted through the chamber, copper was evaporated 3.5 cm downstream from the nozzle using 2.7 kW of e-beam power. Observation of the vapour stream for the 0.50 slm gas flow showed that the vapour distribution approached the $\cos^n\theta$ vapour distribution normally observed directly above high-vacuum e-beam-heated vapour sources (e.g. Fig. 1a)). As the gas flow was increased to 5.0 slm, interaction between the carrier gas flux and the vapour stream redirected the vapour in a broad swathe towards the substrate (Fig. 4a). At much higher gas flows (50 slm), the vapour was confined to the bottom third of the extended fast flow region (Fig.

4b).This set of conditions also revealed that under high gas flow conditions, a significant portion of the ionized vapor stream was redirected parallel to the substrate along with the carrier gas stream. In summary, these experiments revealed that for a constant Mach number, increases in chamber pressure (i.e. carrier gas flux) led to increased vapor transport along, as opposed to across, carrier gas flow lines. Under select conditions however, such as those illustrated in Fig. 4 a), it appears that many of the vapour atoms can be entrained in the core of the carrier gas stream.

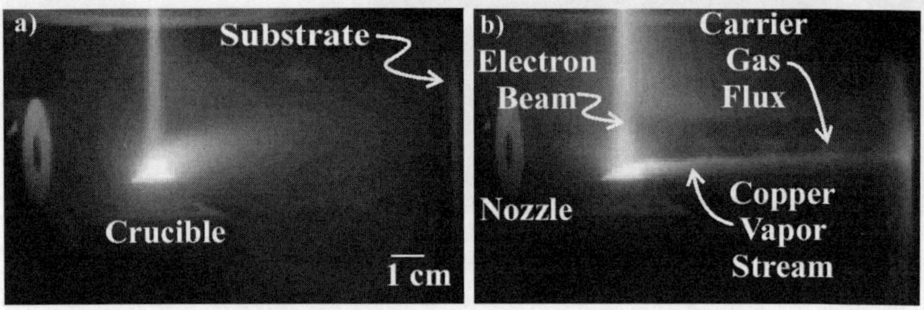

Figure 4 Vapour entrainment into the carrier gas flux. a) $P_d = 0.37$ millibar, Mach 1.7, Carrier gas flux = 5 slm b) $P_d = 2.9$ millibar, Mach 1.7, Carrier gas flux = 50 slm.

The second process variable which was investigated was the Mach number, at constant gas flux (slm) and constant chamber pressure (millibar). To avoid variations in carrier gas/vapour stream interaction due to changes in nozzle diameter, Mach number variations were confined to those possible with a single nozzle diameter of 1.27 cm. The crucible and copper rodstock were positioned 3.5 cm downstream of the nozzle, 2.7 kW of e-beam power were used for evaporation, and 20 slm of helium carrier gas were passed through the system. Mach numbers were varied from 1.5 to 1.9. The remarkable feature of this study was the lack of visible difference in the vapour trajectories [30]. In all cases the flow looked similar to Fig. 4. It should be noted that the minimal changes in vapour trajectories observed are for Mach number changes of only 15 - 20% in total magnitude. In contrast, the carrier gas flux variations of the previous set of experiments represent changes of two orders of magnitude.

The third process variable investigated was the type of carrier gas. Although changing the carrier gas from helium to argon modifies the carrier gas stream shock structure only a little [33], the difference in atomic mass of the two gasses and the different stream velocities generated by identical Mach numbers are likely to affect vapour atom trajectories during transport to the substrate. (Use equation (2) to determine the distinctly different velocities of helium and argon gas flows for identical pressure ratios.) To explore the effect of variations in the type of carrier gas and vapour flux, two sets of visualisation experiments were undertaken in which copper evaporation was performed using 2.1, 4.2, and 6.3 kW of beam power while allowing 25 slm of helium or 5 slm of argon to flow into the system. The helium and argon experiments revealed a similar ability to redirect the vapour flows although the absolute flow of gas in the two sets of experiments was dramatically different. For a given set of gas flow conditions, increasing the evaporation rate decreased vapour focus and increased the ability of the vapour stream to diffuse across or around carrier gas streamlines (Fig. 5). This study suggests that, for a given carrier gas Mach number, vapour atom composition, and carrier gas flux, heavier (but slower) argon atoms redirect vapour atoms much more quickly than do helium atoms. The implications for the overall vapour deposition process are still unclear.

Fig. 6 summarises the characteristic structures and the interaction of the gas and vapour flows in the Directed Vapour Deposition system.

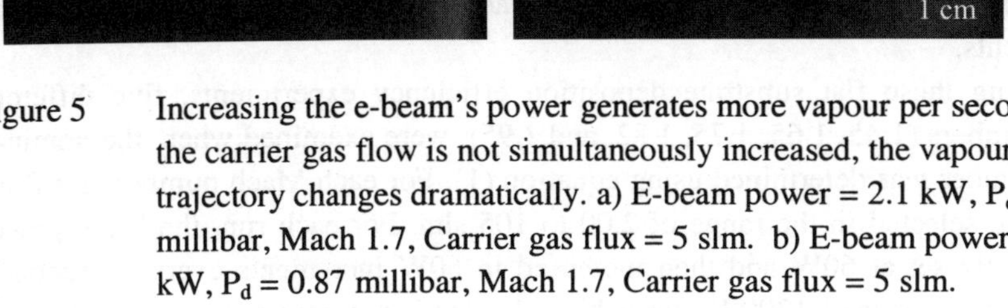

Figure 5 Increasing the e-beam's power generates more vapour per second. If the carrier gas flow is not simultaneously increased, the vapour flux trajectory changes dramatically. a) E-beam power = 2.1 kW, $P_d = 0.75$ millibar, Mach 1.7, Carrier gas flux = 5 slm. b) E-beam power = 6.3 kW, $P_d = 0.87$ millibar, Mach 1.7, Carrier gas flux = 5 slm.

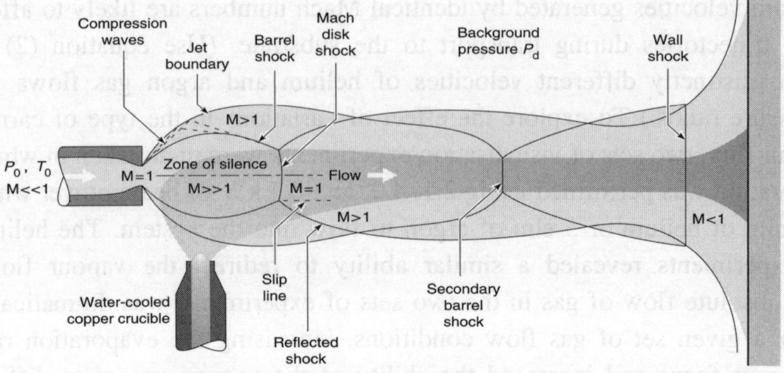

Figure 6 Continuum free-jet expansion combined with vapour entrainment.

3.4. Deposition efficiency at constant e-beam power

Following the visualization experiments, a set of deposition efficiency experiments were performed in order to understand the material utilisation efficiency characteristics of the technology during deposition on to flat surfaces. A similar deposition efficiency study of deposition of material on to small-diameter fibre substrates has been reported elsewhere [29]. An average efficiency of deposition for each individual run was determined by weighing a 1.27 cm diameter copper evaporant source and a flat 10.1 cm by 10.1 cm glass deposition substrate before and after each deposition run. Get tered (purified) helium carrier gas flows were conducted into the processing chamber through a 1.27 cm diameter orifice nozzle. E-beam scanning across the evaporant surface was not employed in this set of experiments.

Using these flat substrate deposition efficiency experiments, five different Mach numbers (1.45, 1.65, 1.75, 1.82, and 1.95) were examined where the nominal Mach number was determined using equation (1). For each Mach number, gas flow rates were selected in the range of 2.00 to 105 slm. For each run, the beam power was initially set at 60W and then increased in 60W increments every 30 seconds until a beam power of 1200W was achieved. This beam power was maintained for ten minutes before being reduced immediately to 60W for thirty seconds and then shut off. The entire cycle took 20 minutes. In these experiments, the centre of the copper rod was 3.5 cm from the nozzle and 9.3 cm from the substrate. Fig. 7 shows the dependence of deposition efficiency upon chamber pressure for various Mach numbers.

As the chamber pressure increased, the deposition efficiency rose quickly before reaching a maximum and then decreased gradually with increasing chamber pressure. To the left of the maximum deposition efficiencies, higher Mach numbers generally yielded higher deposition efficiencies, however to the right of the peak efficiency this trend was reversed, with lower Mach numbers generating higher deposition efficiencies. The specific data, and additional experimental details, are available elsewhere [30].

The deposition efficiency variations with chamber pressure reported in Fig. 7 can be partially reconciled by comparing them with the earlier flow visualizations which showed that when the carrier gas flux or Mach number is low, a significant portion of the vapour diffuses completely across the main carrier gas flow and is never directed toward the substrate. Indeed, for a Mach number/carrier gas flux combination close to zero, the vapour distribution must return to the $\cos^n\theta$ distribution observed in high vacuum e-beam systems (Fig. 1a)). In such a case, a small fraction of the flux would reach a substrate positioned as in these experiments (Fig. 2).

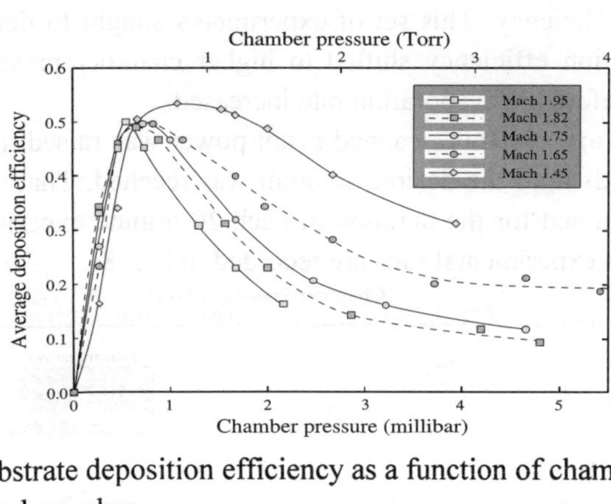

Figure 7 Flat substrate deposition efficiency as a function of chamber pressure and Mach number.

As the carrier gas flux increases, the vapour diffuses into the centre of the carrier gas jet, is transported to the substrate, contacts the substrate, and forms a deposit near the impact point of the jet on the substrate. Once the carrier gas flux increases beyond a certain point however, the vapour is confined to the bottom edge of the carrier gas jet (Fig. 4), is slowed down by the wall shock directly in front of the substrate, and is then deflected into the wall jet parallel to the substrate. Once in the wall jet some vapour atoms will diffuse towards the substrate surface and will deposit, but a significant portion is also retained in the flow and transported out of the chamber, resulting in a decrease in average deposition efficiency as measured here.

The Mach number variation visualisations do not provide significant insight into the Mach number dependence of deposition efficiency as recorded in Fig. 7. It is possible that variation of the exact size and location of the secondary shock structure as recorded in Fig. 3 influences the probability that an entrained vapour atom will reach the substrate.

From the results of this section, it certainly appears that variations in carrier gas conditions can generate wide differences in vapour deposition characteristics, in this case deposition efficiency and thus deposition rate but the implications of these results for vapour deposition quality and flexibility are unclear.

3.5. Deposition efficiency under constant gas flow conditions

To shed additional light upon DVD vapour transport, experiments were undertaken at a fixed Mach number (M = 1.75) but with different levels of scanned e-beam power (2.4, 3.6, 3.9, and 5.4 kW), to examine the effect of evaporation rate variation on deposition efficiency. This set of experiments sought to determine whether the peak in deposition efficiency shifted to higher chamber pressures as the e-beam power and therefore the evaporation rate increased.

For these experiments, scanned beam power was raised in 300 W increments every 20 seconds until the desired setpoint was reached. That setpoint power level was then maintained for the duration of each 20 minute experiment. The results of the beam power experimental runs are recorded in Fig. 8.

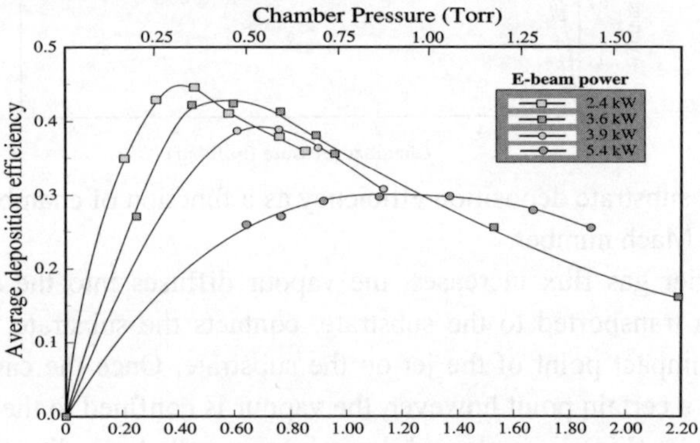

Figure 8 The effect of e-beam power upon deposition efficiency.

Prior to conducting these experiments, the e-beam gun was raised to accommodate a set of e-beam scanning coils at the bottom of the gun. This system modification increased the separation between the gun and crucible from 2.50 cm to 8.80 cm. The increased beam propagation distance in the chamber relative to the original flat substrate deposition study resulted in greater e-beam energy losses in the chamber gas. Because of the introduction of beam scanning and a different distance from gun to source, the results generated in this section come from a distinctly different experimental configuration from that explored in the previous section. While the trends observed in this section and the last should apply to both system configurations, the exact deposition efficiency results can be compared only if the effect of system reconfiguration is considered.

Three distinct trends can be observed in the data of Fig. 8. First, as the beam power increases, the peak of the deposition efficiency curve moves to higher pressures; secondly as the beam power increases, the maximum deposition efficiency of the peak decreases, and thirdly as the beam power increases, the deposition efficiency curve broadens, forming a less distinct peak.

The first trend can be explained quite easily in the light of the visualisations of Fig. 5. Higher beam powers clearly allow the vapour stream to diffuse through and around the primary (fast flow) portion of the gas flow for a given set of carrier gas flow conditions: thus at low flows and high beam powers, the vast majority of the vapour stream passes through the gas jet and over the top of the flat substrate. Only as the gas flow is increased does this higher density vapour flux get redirected towards the substrate by the carrier gas flow. The second and third trends appear to result from the competition between the positive redirection of the higher density carrier gas flow and the negative redirection of the higher density wall jet. Once there is enough carrier gas momentum to redirect the high density vapour flux towards the substrate, there is also enough carrier gas present to direct the vapour into the wall jet before it has a chance to contact the substrate and therefore, even at peak efficiency, the deposition efficiencies of all high-power deposition runs are lower than their low power/lower gas flow counterparts because of the influence of the wall jet.

4. Model-Based Study of "Directed Vapour Deposition"

The carrier gas stream/vapour interactions and the deposition efficiencies of directed vapour deposition have been shown to be sensitive functions of the flow conditions in the system. It is clear that entrainment of e-beam vaporised atoms in a carrier gas stream can modify the inherent velocity vectors of electron beam evaporant. To explore the origins and significance of these effects in more depth, a detailed model for vapour transport in the low vacuum gas stream environment of DVD has been developed.

Researchers have three possible tools for the investigation of phenomena associated with vapour transport: solution of the Navier-Stokes equations, solution of the Boltzmann equation, or direct physical simulation of the gas flow. As explained in [30], direct physical simulation of the gas flow has been selected as the preferred method of analysis in DVD. To track individual vapour atoms one at a time through the DVD process chamber, the Direct Simulation Monte Carlo (DSMC) and bimolecular collision theory (BCT) methods are employed here. Initially, the DSMC method of G. A. Bird [39] was used to calculate the velocity, pressure, and temperature data for the carrier gas flow at specified grid points throughout the modelled region. Once this carrier gas flow data was determined, it was used as the input to a BCT based model that calculated sequentially the mean free path and velocity vector of an individual vapour atom, flowing one at a time through the carrier gas flow (Fig. 9). Vapour atom tracking was performed separately from carrier gas modelling due to limitations of the particular DSMC code employed. In principle, the entire simulation could be performed within a fully developed DSMC code. Ideally an entire flux of vapour atoms would be tracked simultaneously, leaving the crucible and interacting with the carrier gas, however, the models employed here did not allow such a concurrent simulation to be performed. By confining simulations to one vapour atom at a time, it was not possible to investigate evaporation rate phenomena similar to those shown in Fig. 8.

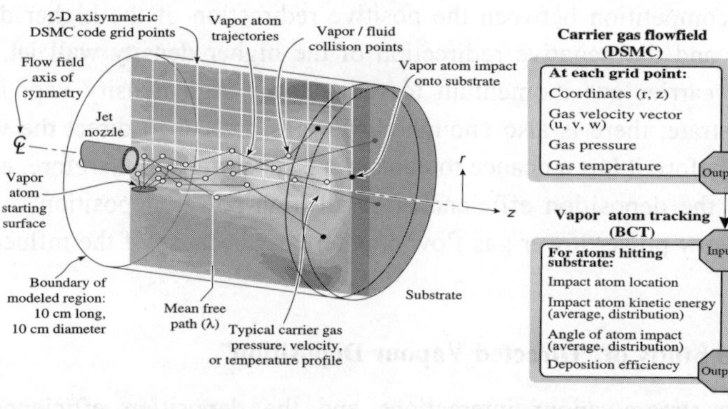

Figure 9 Vapour transport modelling in a low vacuum environment can be performed by discrete atom techniques such as direct simulation Monte Carlo (DSMC) and bimolecular collision theory (BCT).

Nonetheless, the modelling results reported in this chapter yield additional qualitative and quantitative insights into the effect of the carrier gas stream upon individual vapour atom direction and kinetic energy. This study provides new insight into the utility of low or medium vacuum carrier gas stream material processing. When combined with emerging microstructure models [40 - 44], these results can enhance understanding of relationships between the five steps of vapour deposition and final film structure. The study could also provide ways to enhance technology performance in future low and medium vacuum vapour deposition systems.

4.1. Model description

DSMC modeling of the carrier gas flow required generation of a grid of points at which the pressure (P_c), velocity (U_c), and temperature (T_c) of the flow were recorded. The length and radial dimensions of the gridd region were based on the DVD processing space employed for the experimental studies but, to simplify the modeling process, geometric asymmetries such as the crucible were removed from the modelling configuration. This allowed a two-dimensional axisymmetric flow volume to be modelled using a rectangular mesh with a line of symmetry about the flow's central (z) axis (Fig. 9). For all simulations, the spacing between adjacent gridpoints followed Bird's stipulation that cell dimensions, in flow regions with large macroscopic gradients, should be approximately one-third the local mean free path, and time steps over which molecular motion and collisions are uncoupled conformed to Bird's suggestion that they be much less than the local mean collision time [39]. A greater density of gridpoints was placed along the main carrier gas flow line to capture the significant changes in pressure, velocity, and temperature when the flow accelerated out of the nozzle before slowing as it "sensed" the presence of the substrate and changed direction parallel to the substrate surface (e.g. Fig. 4).

All simulations employed a nozzle radius of approximately 6.35 mm. Fluid dynamics research has shown that for transonic ($M = 1$) or supersonic flow ($M > 1$), the velocity, pressure, and temperature conditions at the exit of a nozzle (i.e. the smallest cross-section through which carrier gas flows) always correspond to "choked" or Mach 1.0 conditions. As a result, one-dimensional isentropic[4] flow calculations of a compressible[5] fluid [32, 45] can be used to estimate the initial carrier

---

[4]isentropic - entropy changes to not occur in the flow.

[5]compressible - density variations within the flow are nonnegligible.

gas pressure, velocity, and temperature conditions at the nozzle opening [30]. The validity of employing isentropic theory to establish initial conditions at the nozzle has been well established [46].

Once the DSMC code had been used to establish the carrier gas flowfield, a BCT-based code specially developed for this work was used to simulate vapour atom collisions with individual carrier gas atoms at intervals determined from mean free path calculations [30]. Once an individual vapour atom was released from the surface of the crucible (Fig. 9), it travelled along its constant velocity vector until it collided with a carrier gas atom. Each collision was treated as an elastic, momentum-transferring event which changed the velocity vector of the vapour atom. Each vapour atom underwent multiple collisions until it left the modelled volume or was deposited on the substrate surface. If it was deposited on the substrate surface, its final position and depositing velocity vector were recorded [30].

4.2 Simulation of DVD conditions

DVD modelling runs were undertaken at three different Mach numbers (1.45, 1.75, and 1.95) and eight different chamber pressures (1.333, 4.000, 6.665, 13.33, 26.66, 53.32, 93.31, and 186.6 Pa) to generate detailed insight into the vapour deposition characteristics of the technology. For these computations a helium flowfield was simulated first using Bird's DSMC code. Then individual copper vapour atoms were introduced into the flow one at a time, and followed from the vapour source to the substrate, or to the edge of the modelled flowfield volume, using the BCT code. The centre of the vapour source was located at x = 0.0 cm, y = -0.635 cm, and z = 3.5 cm, given an origin at the centre of the nozzle throat and a total modelled region length of 10 cm, radius 5 cm. (See Fig. 9 for x, y, and z orientation.) For atoms contacting the substrate, a sticking coefficient of one was assumed, and the atom's velocity vector and position at the time of impact were recorded.

As a first investigation of the DVD model, a set of illustrative figures was created in which ten atomic trajectories from crucible source to substrate were superimposed on the helium carrier gas flowfield Mach number (e.g., Fig. 10). The figures represent a two dimensional vertical slice through the cylindrical modelled volume and, although the vapour-emitting surface is shown in these figures, it was not included as a geometric entity in the actual simulations.

First examination of the results shows that the model predicts similar vapour redirection trends to those observed experimentally (Fig. 4). At low chamber pressures and gas flows the simulated vapour atoms received only minimal direction towards the substrate. This appears to have resulted from the minimal redirection imparted by the rarefied carrier gas flow. At higher chamber pressures (Fig. 10), the vapour stream redirection towards the substrate is predicted to be much more pronounced, until the vapour reaches the vicinity of the substrate: once the vapor atoms approach the substrate they begin to wander and travel parallel to the substrate, under the influence of the wall jet.

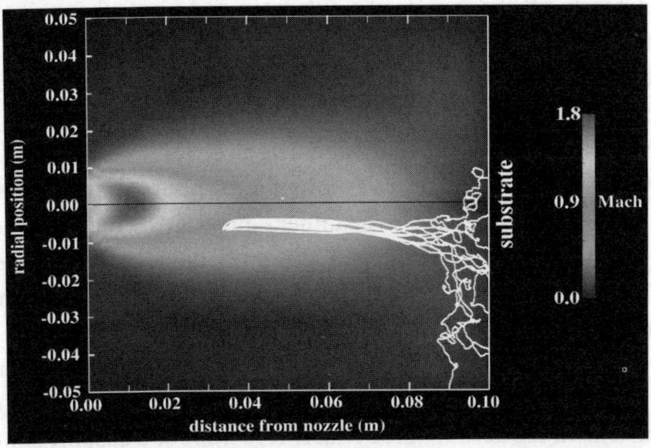

Figure 10 For a background chamber pressure of 130 microbars and a nominal Mach number of 1.75, the model suggests rapid vapour atom redirection towards the substrate and high deposition efficiency.

The model strongly suggests that the difference in vapour atom behavior in the three visualized simulations is the result of differences in the vapour atom mean free path and in the persistence of the fast flow region of the carrier gas stream. At the lowest chamber pressure (13 microbar) and a nominal Mach number of 1.75, the calculated copper mean free path was 450 - 650 μm. For the intermediate and high pressure Mach 1.75 conditions visualised (130 microbar and 1.9 millibar respectively), the calculated mean free paths of the vapour atoms were 30-60 μm and 2-4 μm. Remembering that the separation between vapour source and substrate is 6.5 cm, a first order calculation suggests that vapuor atoms in the lowest pressure simulation undergo 100-200 collisions with background gas atoms prior to reaching the substrate or leaving the modelled volume. At the highest simulated pressure, a similar calculation suggests that vapour atoms undergo as many as 20,000 collisions with background gas atoms prior to reaching the deposition surface or model boundary.

The variation in vapour atom redirection towards the substrate also appears to have been caused by the distinctly different persistence of the fast-flow region of the carrier gas stream. In the low chamber pressure simulation, the model showed that the fast flow region dissipated after traversing only about half the distance from nozzle to substrate. In the higher chamber pressure simulations, the persistence of the fast flow region increased so that at the highest simulation pressure, the fast flow region reached nearly all the way to the substrate and was subsequently redirected into a moderate velocity wall jet (e.g. Fig. 4). Although Fig. 3 hints at such differences in the persistence of the fast flow region, the model has provided far more information about the specific velocity of the carrier gas stream.

While the theoretically predicted vapour redirection trend is similar to experimental observations, the simulations show the vapour atoms being redirected towards the substrate at lower chamber pressures than those observed experimentally. Study of the experimental system behaviour and the predictions of the model [30], indicates that this difference probably results from the model simulation of one vapour atom at a time rather than a flux of atoms.

Additional insight into DVD was gained by plotting the average kinetic energy for twenty vapour atoms as a function of z distance from the nozzle (Fig. 11). The results shown are for the three Mach 1.75 vapour transport simulations just described. To generate the results, the energy of the atoms was sampled at 1mm intervals during transport from source to substrate. In the case of the 13 microbar simulation, an initial rapid energy loss led to low energy vapour transport all the way to the substrate at approximately thermal levels ($E_{thermal} = 0.05$ eV given $T_{He} = 275$ K). This rapid drop-off reflects the fact that, in the low chamber pressure conditions, the carrier gas fast flow region dissipates at about the point where the carrier gas stream reaches the crucible. For the remainder of the distance to the substrate, the flow moves towards the substrate at low velocities.

For the intermediate chamber pressure visualisation (130 microbar), collisions with vapour atoms dissipated the initial vapour atom kinetic energy perpendicular to the carrier gas flow. Continued collisions with the more persistent fast flow region of the carrier jet (Fig. 10) led to energy gain directed toward the substrate. The effect of the wall jet on vapour atom energy was observed as a temporary kinetic energy increase just in front of the substrate as the vapour atoms were accelerated parallel to the flat deposition surface (e.g. Fig. 4). Finally gas/vapour collisions in the boundary layer directly in front of the substrate led to low energy deposition only slightly above background gas thermal levels and well below the deposition energies of high vacuum e-beam systems (approximately 0.2 eV).

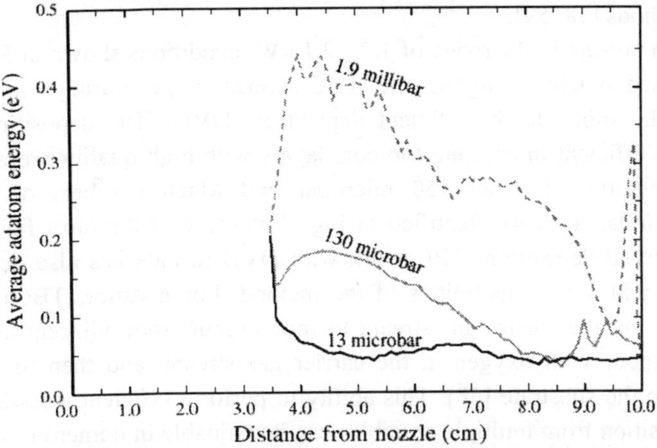

Figure 11 During transport, the energy of vapour atoms under the different flow conditions vary significantly; at the time of deposition however, the energies are remarkably similar at about 0.05 eV.

For the highest chamber pressure simulation (1.9 millibar), the longer, denser fast flow region at this pressure level was able to double the initial vapour atom energy. However, as the velocity of the carrier gas decreased, the velocity and energy of the vapour atoms dropped as well. A slight, temporary rise in vapour atom energy in the vicinity of one of the secondary shock structures was observed followed by a substantial increase in vapour atom energy very close to the substrate (>0.2 eV) as the vapour atoms were turned and accelerated into the wall jet. However, once the vapour atoms diffused through this jet and into the boundary region directly in front of the substrate, they dropped rapidly back towards local thermal levels as a result of multiple collisions with the "dense" low velocity carrier gas.

5. Application of DVD to the Creation of Thermal Barrier Coatings

Based on the studies presented here, application-specific DVD research has been guided towards formation of the top coat layers in thermal barrier coatings (TBCs). TBC systems are composed of a multilayer coating structure that includes a porous columnar ceramic top coat (Fig. 12). Thornton, Zhou, Wadley, and others have demonstrated that porous, columnar atomic structures occur during low energy deposition [53-55]. Initial work with DVD confirms that the method creates porous, columnar top coats for TBCs with properties equal to or better than those produced by other methods [58, 59].

e-beam powers in the range of 1.2 - 2.1 kW, conditions shown in Fig. 5 and 8, have been used to achieve high-quality TBC deposition, generating a vapour stream focused at the substrate for efficient deposition. DVD TBC deposition has also proved to be efficient in creating top coat layers with high quality properties using chamber pressures of about 150 microbar and Mach numbers of about 2.1, conditions similar to those identified in Fig. 7 as some of the most favourable for efficient material deposition. TBC work with DVD to date has also demonstrated some of the additional capabilities of the method. For instance, TBCs have been created by using the carrier gas stream to mix vapour from adjacent crucibles, to react that vapour with oxygen in the carrier gas stream, and then to deposit the product on to the substrate [58]. This ability to perform efficient, possibly reactive vapour deposition from multiple crucibles can be valuable in numerous applications requiring careful control of atomic composition.

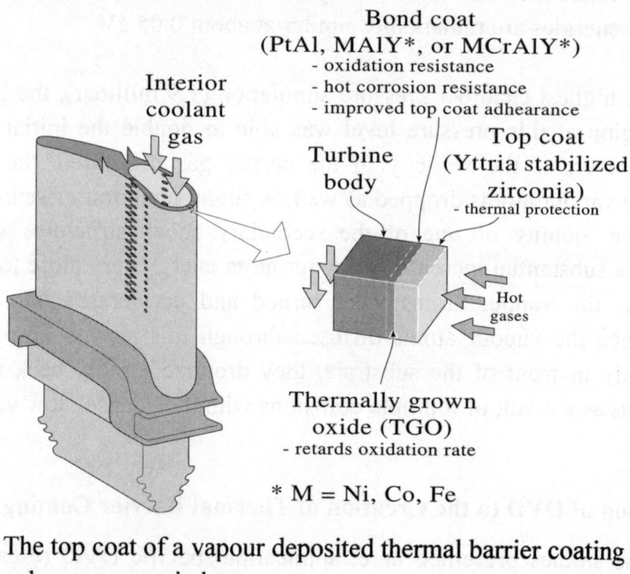

Figure 12 The top coat of a vapour deposited thermal barrier coating is a porous, columnar ceramic layer.

6. Enhancing "Directed Vapour Deposition"

The application of directed vapor deposition to the creation of thermal barrier coatings demonstrates the ability of this new technology to synthesise useful materials. At the outset of DVD development however, one of the goals was to create a processing technology with maximum flexibility for the creation of a wide range of atomic structures and compositions. The initial DVD results presented above suggest an important but limited range of atomic structure and composition control. nonetheless, the need for process flexibility during vapour deposition remains, and thus the question arises: can initial experimental and model-based understanding of DVD be used to develop a more flexible second generation of the technology?

Examination of DVD results suggests that an important limitation upon flexibility results from a system configuration which uses the gas jet to turn the vapour stream through 90° after it leaves the crucible source. Turning the vapour stream limits the ability of the system to operate at low gas flow rates or high evaporation rates. Under low gas flow or high evaporation rate conditions, vapour travels up towards the e-beam gun and over the top of the substrate (e.g. Fig. 5b). If the carrier gas flux is increased to provide enhanced vapour redirection to the substrate, deposition efficiency decreases as the magnitude of the wall jet grows, and clustering of vapour atoms appears to occur [30]. Because so many carrier gas atoms are needed to turn the vapour towards the substrate, the energy and direction of the vapour atoms is determined by carrier gas atoms in their immediate vicinity. Thus, for conditions at which significant numbers of atoms are directed to the substrate, the heavier vapour atoms cannot use their inertia to "punch through" the surrounding carrier gas near the substrate and reach the substrate for high efficiency and possibly high energy deposition.

Based upon these observations of DVD system behaviour, Fig. 13 shows a suggested system reconfiguration which could expand the useful range of DVD technology at reduced gas flows and elevated evaporation rates. Using the DSMC + BCT models described in section 1.4, it is possible to study the behaviour of this proposed reconfiguration prior to construction of the actual tool. While the DSMC + BCT modelling environment cannot be used to investigate evaporation rate effects in a new system, it has been found that this new configuration enhances atomic deposition efficiency and energy.

In the reconfigured DVD system envisioned in Fig. 13, the diameter of the carrier gas nozzle has been enlarged to accommodate the source material, crucible, and gas flow within the throat of the nozzle while also allowing the scanned e-beam to reach each of the separate crucible sources. This larger-diameter nozzle should have an advantageous effect on the gas flow characteristics of the system. The length of the supersonic flow region is known to increase with increasing nozzle diameter, bringing high velocity carrier gas and vapour atoms closer to the substrate and decreasing the distance that energetic vapour atoms must travel through low velocity carrier gas before depositing on to the substrate.

To examine the vapour deposition characteristics of this system reconfiguration, a calculation was run in which the chamber pressure was defined as 13 microbars, and the upstream pressure at the model boundary inside the mixing chamber was defined as 160 microbars. For this simulation the nozzle diameter was 2.20 cm and the copper evaporation surface of the crucible was positioned 0.75 cm down inside the nozzle lip. The crucible was configured to allow evaporation from four neighboring source pools each 3.175 mm in diameter (Fig. 13), an arrangement designed to facilitate mixing of adjacent vapour streams containing materials with significantly different vapour pressures (e.g. yttria-stabilized ceria). For this simulation, evaporation occurred from an individual source pool centred at x = 0.0 cm, y = 0.47625 cm.

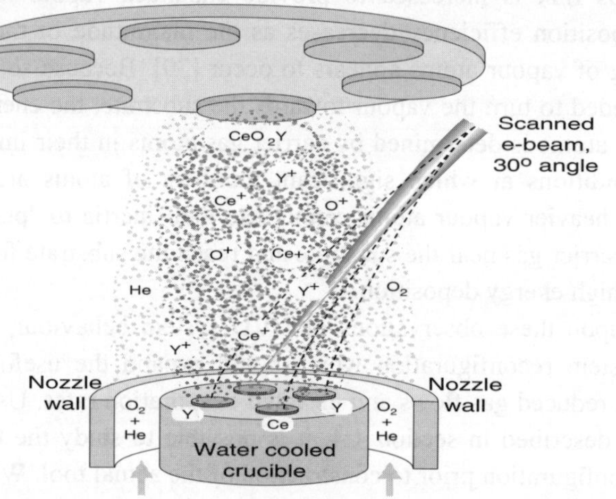

Figure 13 Shown here reactively depositing a TBC top coat layer, this system would use the carrier gas more efficiently, focusing the vapour without having to turn it through 90° towards the substrate.

Study of this simulation (Fig. 14) indicates that the primary advantage of the reconfiguration is a significant increase in deposition efficiency over that of the original DVD system configuration and over that of a conventional e-beam system. The model predicts a deposition efficiency of 83.6% for the simulated conditions. For the same geometrical system configuration but with no carrier gas flow (i.e. high vacuum e-beam evaporation of material), the model predicts a deposition efficiency of 66.5%. Clearly the carrier gas flow performs its intended role, focusing vapour on to the substrate.

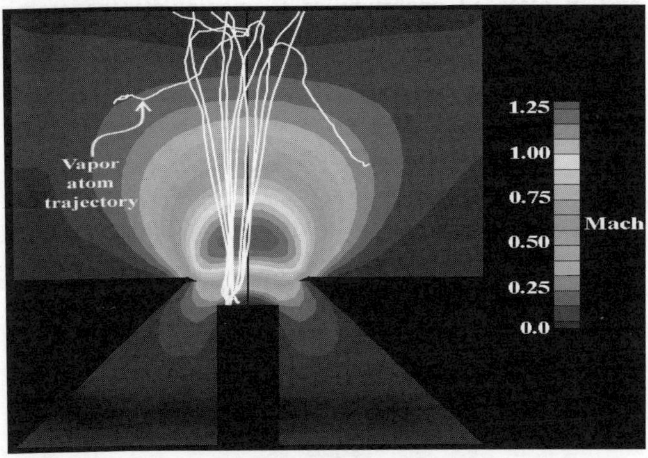

Figure 14 DSMC + BCT simulation of vapour transport in the reconfigured DVD system reveals significantly enhanced vapour deposition efficiency.

Interestingly, the simulation results also indicate that the energy of vapour atom deposition in this reconfigured system is still below initial thermal levels (0.2 eV). The model suggests that the average deposition energy will be about 0.06 eV even though the fast flow region of the carrier gas jet expansion is larger than in the original DVD system. Apparently the mean free path of vapour atoms at this chamber pressure, about 450 - 650 μm, is still short enough to slow the vapour atoms to local, background gas thermal levels. To increase the energy of atoms arriving at the substrate, either the gas pumping characteristics of the technology need to be enhanced or another means of enhancing vapour atom energy needs to be introduced. Following careful consideration of the different available possibilities for enhancing vapour atom energy, it was decided that enhanced atomic structure control could probably be introduced into the system through plasma activation of the DVD gas and vapour stream (Fig. 15).

Although DSMC models can be used to model plasma systems, the particular code employed in this study was not equipped for such investigations. Thus, the decision to introduce a plasma unit into the DVD system was based on careful consultation with experienced researchers at the Fraunhofer Institute for Electron Beam and Plasm Technology (FEP in Dresden, Germany). Previous and ongoing work at FEP strongly suggested that plasma activation in different vapour deposition system afforded opportunities to manipulate the atomic structure of growing films [60].

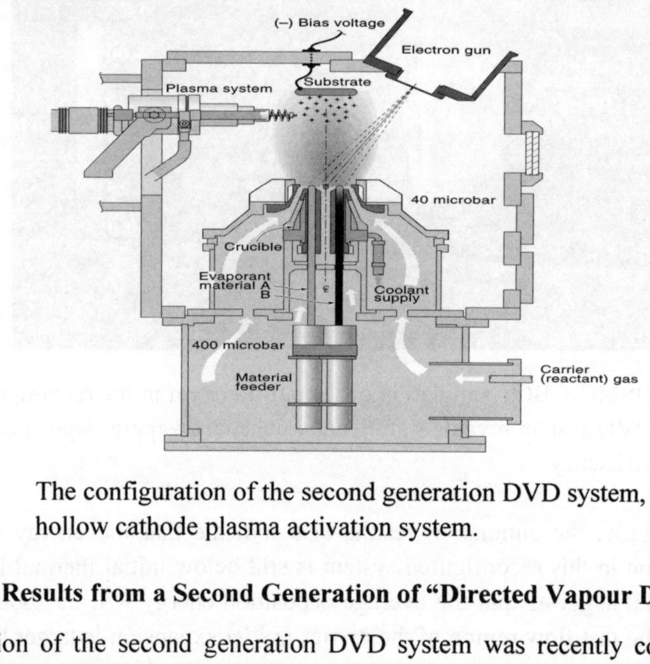

Figure 15 The configuration of the second generation DVD system, including a hollow cathode plasma activation system.

7. Initial Results from a Second Generation of "Directed Vapour Deposition"

Construction of the second generation DVD system was recently completed and preliminary experiments have been performed to verify the operation of the new technology. Fig. 16a shows the ability of the new system to generate a tightly focused carrier gas and vapour stream while Fig. 16b illustrates introduction of hollow cathode plasma activation into the system. Although these pictures clearly show the basic material transport characteristics of the new system configuration, they provide no insight into the atomic structure and composition synthesis capabilities of the second generation DVD system.

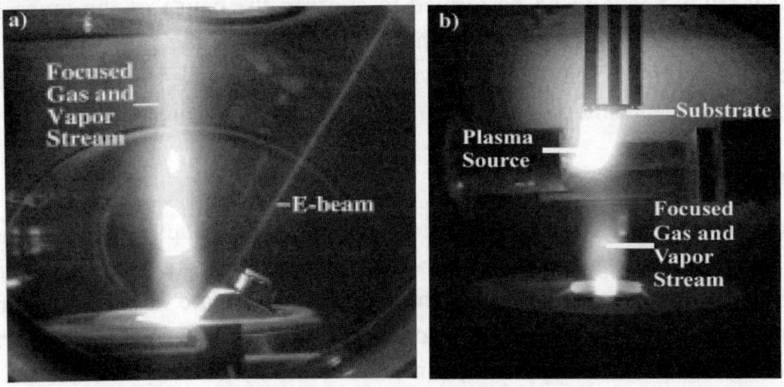

Figure 16 Initial experimental work with the second generation DVD system shows the ability of the system to focus and plasma activate the vapour stream.

To perform an initial evaluation of the material synthesis capability of the new DVD system, two experiments were performed. In both experiments a 1.27 cm diameter rod of zirconium was placed in a single rod crucible inside the nozzle of the new system. In both experiments 4.0 slm of argon and 0.150 slm of oxygen were passed through the nozzle to focus and react with the zirconium evaporated from the crucible using 1.75 kW of e-beam power. These process conditions generated a process chamber pressure of approximately 140 microbar and an upstream, mixing chamber pressure of 1.0 millibar. These conditions correspond to a nominal Mach number of 1.9. For the first experiment, no plasma activation was employed at any point during the fifteen minute long deposition run. Plasma activation of the carrier gas and vapor stream was performed during a second 15 minute deposition run. This second deposition experiment introduced slightly more argon into the system (1.03 slm) to maintain the plasma generated by 40 A of hollow cathode current.

Following the two deposition runs, x-ray analysis of both samples was performed to study the crystallography of the samples. The results are displayed in Fig. 17. Without plasma activation or substrate heating, the second generation DVD system (DVD II) produced an amorphous zirconia deposit. In contrast plasma activation and no substrate heating generated a crystalline, textured deposit. The crystallinity and texture of a heated zirconia sample processed in the original DVD system (DVD I) is shown for comparison. Although no substrate heating was applied to the plasma activated sample, it is estimated that the substrate temperature reached several hundred degrees Celsius during the deposition run. Although additional study needs to be performed to confirm the cause of the amorphous to crystalline transition, it seems likely that the presence of the plasma in the system enhances the energy of the depositing vapour atoms, allowing them to diffuse across the deposition surface more readily into lower energy crystalline sites.

8. Conclusions

The experimental and modelling work described in this chapter and elsewhere has begun to provide insight into the opportunities for electron beam-based material processing in low and medium vacuum. DVD work to date has demonstrated that material processing in a medium and low vacuum environment offers opportunities to affect each step of the vapour deposition process. By passing a carrier gas stream over the surface of an electron-beam-generated melt pool, the inherent vapour creation characteristics of the electron beam system can be modified. DSMC + BCT modelling has shown that during DVD vapour transport, the energy of the vapour atoms in the system is significantly affected as the result of carrier gas/vapour atom collisions. Experimental and modelling results demonstrate the ability of DVD to affect the location of adatom absorption on to the deposition substrate. Through use of a plasma activation unit, it appears that DVD can influence vapour atom diffusion on the deposition surface. Finally, the addition of substrate heating (Fig. 17) also allows bulk diffusion of atoms to create textured crystalline deposits from DVD's low energy deposits.

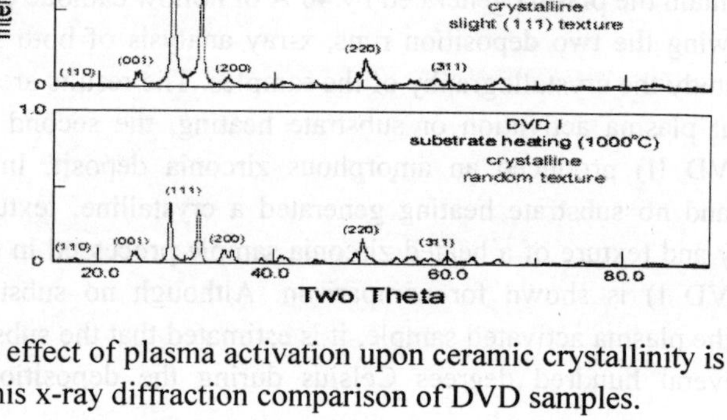

Figure 17 The effect of plasma activation upon ceramic crystallinity is illustrated in this x-ray diffraction comparison of DVD samples.

Although more work must still be done, DVD work so far has begun to demonstrate the flexible material processing characteristics originally envisioned when this new vapour deposition technology was first developed during the early 1990's. Some of the most recent results illustrate the ability of the tool to perform as a high-technology spray painter, a painter that may soon demonstrate the capability of laying down multielement, multilayer, multistructured films on to substrates of many sizes and shapes. Further development of the ability of the system to control atomic structure and composition could lead to the development of a tremendously flexible and valuable coating tool. While the cost of this technology may prevent its use in certain high volume, low margin coating applications, its ability to fabricate sophisticated coatings may lead to its introduction as the preferred method of material creation in other market segments.

Acknowledgments

The authors thank Derek Hass for his contributions to the development of this chapter. They would also like to recognise the support of DARPA (W. Barker, Program Manager), NASA (D. Brewer, Technical Program Monitor), and the AFOSR (W. Jones, Program Manager) during early DVD development. The authors acknowledge the support of the Office of Naval Research (Steve Fishman, Technical Program Monitor) which has allowed the second generation DVD system to be built. Finally, they would like to express sincere appreciation to the scientific staff of the Fraunhofer Institute for Electron Beam and Plasma Technology, particularly Goesta Mattausch and Henry Morgner, for their dedicated efforts which have allowed the second generation DVD system to move from idea to reality.

References

1. Ohring, M. (1992). *The Materials Science of Thin Films*. Academic Press, Inc. New York.
2. Maissel, L. I. and Glang, R. (1970). *Handbook of Thin Film Technology*, McGraw-Hill Book Company. New York.
3. Holland, L. (1956). *Vacuum Deposition of Thin Films*. John Wiley & Sons, New York.
4. Baumgartner, H. (1996). Spraying virtual iron on real engines. *Mechanical Engineering*. 84.

5. Wolf, S. and Tauber, R. N. (1986). *Silicon Processing for the VLSI Era, Vol. 1 - Process Technology.* Lattice Press, Sunset Beach, CA.
6. Anthony, T. C., Brug, J. A., and Zhang, S. (1994). *IEEE Transactions on Magnetics*, **30**(6), 3819.
7. Meier, S. M. and Gupta, D. K. (1994). The evolution of thermal barrier coatings in gas turbine engine applications, *Transactions of the ASME*, **116**, 250.
8. Pindera, M.-J., Williams, T. O., and Arnold, S. M. (1994). Thermoplastic response of metal-matrix composites with homogenized and functionally graded interfaces. *Composites Engineering*, **4**(1), 129.
9. Schiller, S., Heisig, U., and Panzer, S. (1982). *Electron Beam Technology*, John Wiley & Sons, New York.
10. Bunshah, R. F. ed. (1994). *Handbook of Deposition Technologies for Films and Coatings*, 2nd ed. Noyes Publications, Park Ridge, NJ.
11. Ritter, E. (1966). Deposition of oxide films by reactive evaporation, *J. Vac. Sci. Technol.* **3**(4), 225.
12. Bunshah, R. F. and Raghuram, A. C. (1972). Activated reactive evaporation process for high rate deposition of compounds, *J. Vac. Sci. Technol.* 9(6), 1385.
13. Mattox, D. M. (1973). Fundamentals of ion plating. *J. Vac. Sci. Technol.* **10**(1), 47.
14. *Handbook of Thin Film Process Technology*. (1997). ed. D. A. Glocker, S. I. Shah, Institute of Physics Publishing, Philadelphia.
15. Hill, R. J. ed. (1986). *Physical Vapor Deposition*, The BOC Group, Inc. Berkeley, CA.
16. Smith, Jr. H. R., Kennedy, K. and Boericke, F. S. (1970). Metallurgical characteristics of titanium-alloy foil prepared by electron-beam evaporation. *J. Vac. Sci. Technol.* 7(6), S48.
17. Hughes, J. L. (1974). Making alloy foils by electron beam evaporation. *Metals Engineering Quarterly*, 1.
18. McCullough, C., Storer, J., Berzins, L.V. (1995). Manufacture of orthorhombic titanium aluminide composites by PVD methods. in *Recent Advances in Titanium Metal Matrix Composites*. eds. F.H. Froes and J. Storer, TMS, Warrendale, PA, 259.
19. Beale, H. A., Weiler, F., and Bunshah, R. F. (1973). Evaporation variables in gas scattering plating processes. *Proceedings of the Fourth International Conference on Vacuum Metallurgy*. Tokyo, Japan, 238.

20. Hill, D. L. (1994). Jet vapor deposition coating of fibers. *Masters Thesis*, University of Virginia.
21. Nichols, C. A., Rossnagel, S. M., and Hamaguchi, S. (1996). Ionized physical vapor deposition of Cu for high aspect ratio damascene trench fill applications. *J. Vac. Sci. Technol.* **B14**(5), 3270.
22. Halpern, B. L., et al. (1999). H-atom assisted jet vapor deposition of parylene-n polymer. MRS Proceedings, **544**, 153.
23. Guo, X., Ma, T. P., Tamagawa, T., and Halpern, B. L. (1998). High quality ultra-thin TiO_2/Si_3N_4 gate dielectric for giga scale MOS technology. Technical Digest - International Electron Devices Meeting, 377.
24. Schiller, S., Heisig, U., and Goedicke, K. (1975). Alternating ion plating - a method of high-rate ion vapor deposition. *J. Vac. Sci. Technol.* **12**(4), 858.
25. Eastman, J. A., Thompson, L. J., and Marshall, D. J. (1993). Synthesis of nanocrystalline materials by electron beam evaporation. *Nanostructured Materials*, **2**, 377.
26. Halpern, B. L. and Schmitt, J. J. (1994). Multiple jets and moving substrates: Jet Vapor Deposition of multicomponent thin films, *J. Vac. Sci. Technol. A*, **12**(4), 1623.
27. Rechsteiner, R., and Ganiere, J. D. (1981). Calculation of homgeneous nucleation for a source of clusters with thermalizing carrier gas. *Surface Science*, **106**, 125.
28. Marple, V. A. and Willeke, K. (1976). Impactor design. *Atmospheric Environment* **10**, 891.
29. Groves, J. F. and Wadley, H. N. G. (1997). Functionally graded materials synthesis via low vacuum directed vapor deposition. *Composites Engineering, Part B*, **28**(1-2), 57.
30. Groves, J. F. Ph.D. Diss., (1998). http://www.ipm.virginia.edu/process/PVD/Pubs/thesis5.htm, U. of Virginia.
31. Wadley, H. N. G. and Groves, J. F. Directed Vapor Deposition of electron beam evaporant, U. S. Patent No. 5,534,314, July 9, 1996.
32. Fox, R. W. and McDonald, A. T. (1992). *Introduction to Fluid Mechanics*, 4th ed. John Wiley & Sons, New York.
33. *Atomic and Molecular Beam Methods*. (1988). Vol. 1, ed. G. Scoles, Oxford University Press, New York.
34. Adamson, Jr. T. C. and Nicholls, J. A. (1959). On the structure of jets from highly underexpanded nozzles into still air. *J. of the Aerospace Sciences*, **26**(1), 16.

35. Mikami, H. (1982). Transport phenomena in free-jet expansions. *Bulletin of the Research Laboratory for Nuclear Reactors*, 7, 151.
36. Hass, D. D. (1997). University of Virginia, Private communication.
37. Groves, J. F. and Wadley, H. N. G. (1997). Functionally graded materials synthesis via low vacuum Directed Vapor Deposition, *Composites Part B*, **28**(1-2), 57.
38. Shames, I. H. (1982). *Mechanics of Fluids*, 2nd ed. McGraw-Hill Book Company, New York.
39. Bird, G. A. (1994). *Molecular Gas Dynamics and the Direct Simulation of Gas Flows*. Clarendon Press, New York.
40. Yang, Y., Johnson, R. A., and Wadley, H. N. G. (1997). A Monte-Carlo simulation of the physical vapor deposition of nickel. *Acta mater.* **45**(4), 1455.
41. Zhou, X. W., Johnson, R. A., and Wadley, H. N. G. (1997). Vacancy formation during vapor deposition. *Acta mater.* **45**, 4441.
42. Zhou, X. W., Johnson, R. A., and Wadley, H. N. G. (1997). A molecular dynamics study of nickel vapor deposition: temperature, incident angle and adatom energy effects. *Acta mater.* **45**(4), 1513.
43. Zhou, X. W., and Wadley, H. N. G. (1999). Twin formation during the atomic deposition of copper, *Acta mater.* **47**, 1063.
44. Zhou, X. W. and Wadley, H. N. G. (1999). Hyperthermal vapor deposition of copper: athermal and biased diffusion effects, *Surf. Sci.* **431**, 42. Zhou, X. W. and Wadley, H. N. G. (1999). Hyperthermal vapor deposition of copper: reflection and resputtering effects, *Surf. Sci.* **431**, 42.
45. John, J. E. A. (1984). *Gas Dynamics*, 2nd ed. Allyn and Bacon, Inc. Boston.
46. Boyd, I. D., Beattie, D. R., Cappelli, M. A. (1994). Numerical and experimental investigations of low-density supersonic jets of hydrogen. *J. Fluid Mech.* **280**, 41.
47. McDaniel, E. W. (1964). *Collision Phenomena in Ionized Gases*. John Wiley & Sons, Inc. New York.
48. Ziegler, J. F., Biersack, J. P. , and Littmark, U. (1985). *The Stopping and Range of Ions in Solids*. Pergamon Press, New York.
49. Gordon, R. G. and Kim, Y. S. (1972). Theory for the forces between closed-shell atoms and molecules. *J. Chem. Phys.* 56(6), 3122.
50. Kim, Y. S. and Gordon, R. G. (1974). Ion-rare gas interactions on the repulsive part of the potential curves. *J. Chem. Phys.* 60(11), 4323.

51. Kharchenko, V., Tharamel, J., and Dalgarno, A. (1997). Kinetics of thermalization of fast nitrogen atoms beyond the hard sphere approximation. *J. of Atmospheric and Terrestrial Physics*, In Press.
52. World wide web site of the Laboratory of the Physics of Gases and Plasmas, University of South Paris, Orsay, France. Access address http://gaphyor.lpgp.u-psud.fr/
53. Thornton, J. A. (1975). Influence of substrate temperature and deposition rate on structure of thick sputtered Cu coatings. *J. Vac. Sci. Technol.,* **12**(4), 830.
54. Thornton, J. A. (1987). Structure-zone models of thin films, *SPIE vol. 821, Modeling of optical thin films*, 95.
55. Zhou, X. W., Johnson, R. A., and Wadley, H. N. G. (1997). A molecular dynamics study of nickel vapor deposition: temperature, incident angle and adatom energy effects. *Acta mater.* **45**(4), 1513.
56. *Atomic and Molecular Beam Methods*. (1988). Vol. 1, ed. G. Scoles, Oxford University Press, New York.
57. Groves, J. F., et al. (1997). Electron beam directed vapor deposition. *Proc. Electron Beam Melting and Refining*, Bakish Material Corp. 46.
58. Hass, D. D., Parrish, P. A., and Wadley, H. N. G. (1998). Electron beam directed vapor deposition of thermal barrier coatings. *JVST A*, **16**(6), 3396.
59. Hass, D. D., Slifka, A. J., and Wadley, H. N. G. (2000). Low thermal conductivity vapor deposited zirconia microstructures. *Acta mater.*, submitted.
60. Schiller, S., Kirchhoff, V., Schiller, N., and Morgner, H. (2000). PVD coating of plastic webs and sheets with high rates on large areas. *Surface and Coatings Technology*, **125**(1), 354.
61. Johannes, J., Bartel, T. J., Economou, D. J., Lymberopoulos, D. P., and Wise, R. S. (1996). Simulation images from a low-pressure chlorine plasma reactor using DSMC. *IEEE Transactions on Plasma Science*, **24**(1), 127.

Index